# Rudolf Steiner's Contributions to the History and Practice of Agricultural Education

by

**Hilmar Moore**

BIO-DYNAMIC Farming
and Gardening Association, Inc.
P.O. Box 550, Kimberton, PA 19442
U.S.A.

ISBN 0-938250-33-7

printed in U.S.A.



BIO-DYNAMIC Farming
and Gardening Association, Inc.
P.O. Box 550, Kimberton, PA 19442
U.S.A.

## TABLE OF CONTENTS

Preface 1

**Part I: Rudolf Steiner and the Modern Scientific World-View**

Chapter

I. INTRODUCTION 6

II. RUDOLF STEINER'S LIFE AND WORK 16

**Part II: Agriculture and the Development of Human Consciousness**

III. FROM PRE-HISTORY THROUGH GREECE AND ROME 30

IV. MYSTERY CENTERS, MYTHS, AND THINKING 50

V. THROUGH THE MIDDLE AGES TO MODERN TIMES 64

**Part III: Rudolf Steiner's Approach to Agriculture and Education**

VI. BIODYNAMIC AGRICULTURE 84

VII. WALDORF EDUCATION AND ECOLOGICAL VALUES 101

VIII. WALDORF SCHOOLS AND BIODYNAMICS 125

IX. CONCLUSION 137

APPENDICES 147

SELECTED BIBLIOGRAPHY 151

# PREFACE

This book is a revision of the dissertation I submitted as part of a doctoral program at the Union Institute in Cincinnati, Ohio. In it, I tried to bring new ways of thinking about the relationship between education and agriculture, particularly the question: "how can we educate young people in ways that affirm a new view of our relationship to nature and that will be helpful in forming life-affirmative, holistic values."

During the fourteen years that have passed since I wrote it, many changes have occurred in agriculture. Biodynamics has become much more widely known in the United States and, indeed, is now recognized for providing the highest quality of produce and innovative social-economic organizational forms. Biodynamic farmers have taken a leading role in formulating and implementing various certification programs at the state and international levels. Perhaps the greatest change has been the increase of biodynamic farms and market gardens—when I wrote the dissertation in 1977, there were many more biodynamic home gardeners than commercial growers. It is most gratifying that today more growers can work full time in biodynamics, and that the population is becoming more aware of the need to support farms that bring new health and life to the soil which then gives us healthy, life-enhancing produce in return.

Waldorf education was already growing rapidly in 1977. There are now well over 100 schools in the U.S. alone and over 400 throughout the world. Most of these schools attempt to implement Steiner's indications for the use of horticulture in their curriculum—with varying degrees of success. I did not realize the many difficulties that face the implementation of gardening in the Waldorf curriculum. The reasons for this lie in the history of agriculture and education.

Today agriculture is at its lowest ebb in human history. For centuries, human beings depended on the bounty of the earth for their existence and agriculture was a respected profession. Now fewer than two per cent of the population of this nation are directly involved in agricultural production. When the severe agricultural depression of the 1980s devastated farmers, particularly in the middle of the country, the prevalent reaction was "well, they are businessmen like any others, and there are simply too many farmers anyway. They should go out of business and find another occupation."

Such a view is most short sighted and is based on the fallacy that agriculture is like any other job—it is merely another form of industrial activity. Little has been done to help people who want to devote their lives to agriculture; the prospects for professional farmers was, and is, not very good, at least in conventional agricultural ventures. As Americans have turned their backs on farming, this refusal to take responsibility for the land has caused great damage, not only to *our* soil, water, and air, but to the environment of other nations who grow food to export to the U.S. and Europe and who have adopted wholeheartedly the worst aspects of scientific, chemically-based agriculture.

The situation for Waldorf teachers of gardening is not good at present, although one can participate in gardening at Rudolf Steiner College under the expert guidance of Harald Hoven, and extremely motivated students at the Waldorf Institute can find some time to work at the nearby Fellowship Community gardens under Machado Mead. But the reality is that *academic institutions and agriculture do not mix easily*. The academic world and agriculture have lived in vastly different realms for too long, and our present civilization, as discussed above, does not value agriculture.

In Waldorf schools, there is usually too much to be done by too few people and with too limited financial resources to allow a faculty member to devote himself or herself to maintaining a permanent school garden. Often, a person suited for gardening does not have the gift of relating easily with children and gifted teachers have no feeling for gardening. The plants grown in a garden do not adapt their needs well to the limitations of the school year, so the garden gets neglected during the summer. Where teachers make a sincere effort to garden with children, they respond positively, especially the younger ones. Many schools do try to do some gardening, and many schools take the children for visits of a day or a week to a biodynamic farm. Despite good intentions and effort, then, gardening has not found the place in Waldorf schools that it *should* have and, someday, *will* have.

❄ ❄ ❄ ❄ ❄

The problems I saw in agriculture and the attitude toward it fourteen years ago that led me to write this book are still there, but there have been some hopeful developments as well. The Waldorf schools remain perhaps the most successful institutions to develop out of Rudolf Steiner's life and work, and as a parent of two Waldorf students, I have seen how deeply they learn to care for the environment. The garden at the public elementary school here in Dripping Springs, Texas, is certified by the Texas Department of Agriculture as an organic garden! Clearly, there are teachers who know that children must be aided in approaching nature in a new way, and teachers in many schools are searching for ways to do this.

The fact that my dissertation has sold well in xerographic form, published by University Microfilms at an exorbitant price, shows that there are some people who found it helpful in thinking about teaching agriculture and a life-affirming attitude to children. Perhaps most significantly, many people have appreciated the central portion of this work, which employs Rudolf Steiner's concept of the evolution of human consciousness to survey the history of agriculture, and they have found it to be of assistance in forming new ideas about agriculture and its rightful place in human culture and civilization. While I think my introductions to biodynamics and Waldorf education are certainly adequate, it is my research on agriculture and the development of human consciousness that makes the greatest contribution to the existing literature.

Thus, when Rod Shouldice, executive director of the Biodynamic Farming and Gardening Association, gave me the opportunity to revise the thesis into a new book, I reread it and discovered that it may well continue to serve as a guide for teachers, farmers and gardeners, and all those who search for a new way to relate to the earth. It is with this hope that I have undertaken to revise the original work, and it is with heartfelt thanks to those people who seek to take renewed responsibility for the earth—farmers, gardeners, parents and teachers— that I offer this attempt to point to new directions in agriculture and education.

## Purpose and Methodology

This book examines Rudolf Steiner's work in agriculture and education, particularly the historical and descriptive aspects of his theories and methods, and the work in agriculture and pedagogy that has arisen from his indications, both during his lifetime and since his death in 1925. I have attempted to place Steiner's life and work in its historical and cultural context of the nineteenth and twentieth centuries. I have employed Steiner's method of researching the history of ideas from the perspective of the evolution of human consciousness in order to provide a frame of reference for his educational and agricultural work. The work of Ernst Lehrs, Ehrenfried E. Pfeiffer, A.C. Harwood, Guenther Wachsmuth, Francis Edmunds, and others is used to supplement Steiner's work on historical-cultural development, education, and agriculture.

I have synthesized the historical indications in Rudolf Steiner's writings and lectures which directly or by implication can be seen as attitude-forming for the teaching of agricultural methods and ecology, both in modern times and in the past. I investigated Steiner's educational practices specifically as applied to the teaching of agriculture, and I surveyed the work currently done along these lines in various cultural institutions.

The first chapter introduces the concept of the relationship of agriculture to the manifold social and economic problems of today, many of which are related in turn to science and technology. Agriculture increasingly has fallen under the domination of the money-market and technology, in the form of carrying huge debts and the use of chemicals and heavy machinery. Farming is seen today much more as a type of manufacturing and commerce, not as a natural, organic process tied to regional and global ecology. This view is partially a result of the modern scientific world-conception, and I discuss the growing recognition and criticism of such a world-view as a reliable guide to humanity in the larger questions of life. Some evidence follows to show that this world-view seems more and more incapable of providing either nourishing produce, a nourishing environment, or "nourishing" knowledge as a basis for meaningful human actions. I develop the hypothesis that through a study of the history of ideas from the perspective of the evolution of human consciousness, one can discern where and when these problems of knowledge arose. Finally, I examine Goethe's contributions to epistemology and the natural sciences in the light of this evolution.

Goethe's scientific work came to fruition through the efforts of Rudolf Steiner. In Chapter Two, I describe four aspects of the life and work of Steiner: his innovations in epistemology following his study of Goethe; the implications for science of Steiner's theory of knowledge; the development of his psychology and pedagogy; and his indications for a biologically sound agriculture. I have presented these ideas in their historical context in order to lay a foundation for my subsequent exploration of Steiner's contributions to the history and practice of agricultural education.

Chapters Three, Four, and Five consider the development of agriculture from Steiner's conception of history. Perhaps the leading motif that emerges is the fragmentation of the ancient unity of culture—of art, science, and religion; thus science began to lose the imaginative element of artistic feeling and the inspiration derived from religion, and it became ever more tied to an outlook based on rationalistic materialism. I discuss Steiner's research into the ancient temple cultures, and his view of the Mystery centers as the source of humanity's knowledge of and training in spiritual matters, how this knowledge permeated older cultures, and how it became submerged by the scientific-materialistic outlook. In education, Steiner said, an important development was the increasing role of the intellect divorced from art and religion. Where the Greek educator dealt with the whole human being through the gymnasium and academy, and the medieval ideal was the goddess "Natura" surrounded by her seven beautiful maidens representing the seven liberal (liberating) arts, the modern ideal is the "professor" or doctor of philosophy. This meant to Steiner that a *quantity* of knowledge was now the goal, not the development of *abilities* that Greece, Rome, and even the medieval schools had tried to inculcate. I trace the role of Mystery schools in maintaining the old knowledge of the spirit, and how Steiner thought that cultural renewal, including agriculture and education, demands that professional development and self development—outer and inner life—must merge and become complementary.

These chapters can assist the reader in coming to terms with *why* we must find new ways to farm and to assume responsibility for the stewardship of the land, and *why* new principles of education are needed.

Chapters Six and Seven deal with Steiner's work in agriculture and education. Specifically, in the discussion of biodynamics I emphasize the attitude of mind necessary for a thorough comprehension of Steiner's teachings of the earthly and cosmic rhythms. It is this outlook, so different from the reductionist view of the natural sciences, that ties the chapter to the previous section; thus one aspect of Steiner's work is a restatement of the earthly-cosmic-human correspondences once taught in the temple sanctuaries. In Chapter Seven, I outline Steiner's educational ideas, particularly the aspects of the Waldorf curriculum that can engender life-affirmative, holistic values. I compare the demands placed on the biodynamic farmer with those of Waldorf school teachers.

A description of some institutions that employ Steiner's indications in agriculture and education, either to use gardening and farming in education or to teach biodynamics, written with help from Rod Shouldice, comprises Chapter Eight. In the final chapter I offer some concluding comments on the growing activity in Waldorf education and biodynamic agriculture.

I am quite aware of the limitations which the methodology of this endeavor entails. It is an attempt to illuminate several aspects of the work of a most comprehensive thinker who appears to be the most universal mind of our century. Accordingly, I have had to take an interdisciplinary approach, and the result is an amalgam of the history of ideas, cultural history, and, of course, agriculture and education in theory and practice.

Perhaps in my methodological approach I reveal my biases. For I believe that we are living in a crucial time in history, when ecological disaster threatens, when our culture and civilization is decaying and under increasing attack from many directions. The inner core of our cities is devastated by crime and poverty and the blight creeps outward toward the suburbs and into rural areas; the recent statistics and articles on public schools show a decline in nearly all aspects of education, despite unprecedented sums of money being poured into the schools. The scientific world-view paradoxically has brought in a new age of authority in which hypothetical models are promulgated as the only acceptable representations of physical reality, and in which intimations of the spiritual side of things are called abstract, poetic nonsense or worse.

At the same time, the past half-century has seen a rapidly growing interest in spiritual matters, an interest that has spawned all manner of gurus, charlatans, and some genuine leaders, and through which such formerly arcane ideas as reincarnation are now the topic of movies and novels. It seems to me that some basis for the reunification and revitalization of our culture is necessary. In my readings and searchings through the philosophy and psychology of both East and West, it became apparent to me that a new world-conception must be found, one large enough to encompass the wisdom of oriental and western mysticism and philosophy, and also the gains in thinking and observation made by western science. In a variety of areas, including agriculture, education, medicine, and philosophy, the work of Rudolf Steiner has initiated a cultural movement originating from such a world-conception based on his understanding of humanity's relationship to the worlds of spirit and matter.

This book, then, draws from Steiner's vast body of work his indications on the history and practice of agricultural education and forges them into a cohesive narrative. I hope to make Steiner's ideas on this subject accessible to the many people searching for a biologically safe, sustainable agriculture and a life-enhancing education.

## A Note on Footnotes

In this book, I have used the following method to cite works published by the two major firms that publish anthroposophical authors:

1. The Rudolf Steiner Press, London, England is cited as (London: date).
2. The Anthroposophic Press, Hudson, New York is cited as (New York: date).

# PART ONE:

# RUDOLF STEINER AND THE MODERN SCIENTIFIC WORLD-VIEW

## CHAPTER ONE: INTRODUCTION

### Some Aspects of Agriculture

There is hardly an area of life which agriculture does not touch in some way, and many problems arise through its pervasive, all-encompassing nature. For example, current environmentally destructive agricultural practices resulted partly from increased economic demands placed on farmers by the depletion of the farm work force which streamed into the cities. Farmers needed as many labor-saving methods as possible, and these methods led to serious soil damage and to large scale water pollution. While the new industrial workers gained better pay in the cities, the social costs of increased pollution, social alienation in the city, loss of valuable farm lands to encroaching suburbs, and the social devastation of many rural areas due to the urban migrations are only now being computed.[1] Property tax rates which do not differentiate between commercial, residential, and agricultural land have caused many farmers to sell their land to real estate developers,

In 1938, then Secretary of Agriculture Henry A. Wallace wrote:

> The earth is the mother of us all—plants, animals, and men. The phosphorus and calcium of the earth build our skeletons and nervous system. Everything else our bodies need except air and sun comes from the earth.
>
> Nature treats the earth kindly. Man treats her harshly. He over plows the crop land, overgrazes the pastureland, and over cuts the timberland. He destroys millions of acres completely. He pours fertility year after year into the cities, which in turn pour everything they do not use into the rivers and the ocean . . .
>
> The social lesson of soil waste is that no man has the right to destroy even if he does own it fee simple. The soil requires a duty of man which we have been slow to recognize.[2]

In the same year Gove Hambridge wrote:

> A certain man had a fine horse that was his pride and his wealth. One morning he got up early to go out to the stable, and he found it empty. The horse had been stolen. He stayed awake many nights after that thinking what a fool he had been not to put a good stout lock on the door. It would have cost only a couple of dollars and saved his most prized possession. He resolved that he would give better protection to the

next horse he had, but he knew that he would never get one as good as the one he had lost. The United States has been like that about its soil. Within a comparatively short time, water and wind have flayed the skin off the unprotected earth, causing widespread destruction, and we have been forced to realize that this is the result of decades of neglect. The effort to relieve economic depression for farmers has also forced attention to the soil. In the old Roman Empire, all roads led to Rome. In agriculture, all roads lead back to the soil, from which farmers make their livelihood.[3]

In the fifty years since these men spoke out, progress has been made in soil conservation. Yet economic pressures are equally relentless today and result in rural depopulation, a trend to ever larger farms, and poor land management.

In Michigan, a fourth-generation family farm run by Robert Thompson, his father, and two younger sons, carries over $500,000 in debt. He says, "the debt doesn't worry me anymore. It used to. Then I figured that all I was going to make was a living anyhow. The debt is just a fact of farming." Thompson keeps up his interest payments and after a good year, he can pay off some of the principal. Thompson's sons carry a $218,000 debt themselves; the eldest, Mike, says "City people would probably go nuts with a debt like this hanging over them. But in this business you have to stay in debt or Uncle Sam would take it all in taxes."[4]

Karl T. Wright, an agricultural economist, estimated in 1976 that in the next nine years (by 1985), 20,000 family !arms would disappear in Michigan alone, and some seventy American farms are sold each day. His prediction has proved correct. Even moderate-sized farms are being squeezed out by severe economic pressures, not the least of which are the ten billion dollars in government subsidies that go to large farmers. The large corporate farms can afford to take a small profit or a loss on food-growing to offset taxes. The escalating land prices and land taxes hurt the small to moderate-sized farms and benefited the largest ones. Thus the large corporate farm is more "economical" in a financial sense, but in ecological terms this means more insecticides, rural depopulation, and more fertilizer.[5]

In Iowa, farmers raised about fifty bushels of corn per acre in 1950, with negligible fertilizer use. In 1958, they applied 100,000 tons of nitrate fertilizer, and the yield curbed to seventy bushels per acre. By 1965, the application of over 400,000 tons of nitrate yielded ninety-five bushels of corn per acre. The law of diminishing returns is clearly at work: a 400% increase in fertilizer gave only 72% more yield, and the extensive use of petroleum to manufacture fertilizer causes its price to grow ever higher. By 1972, an Iowan farmer needed to produce at least eighty bushels per acre merely to break even, and "this can only be achieved by using nitrogen at levels that are utilized very inefficiently by the crop." The excess nitrogen runs off into rivers, lakes, and into aquifers, causing massive water pollution problems.[6]

Poor farming practices such as these are causing fifty million acres of American farmland to erode rapidly enough to be ruined within twenty-five years, if not stopped. The next quarter-century will see an additional fifty million acres of arable land taken by urban growth. Economic and social pressures have caused American farms to dwindle from 5.6 million in 1950 to 2.8 million in 1975, and farm workers have decreased from 9.9 million to 4.5 million over the same period.

By 1991, fewer than 2% of the population were farmers. On a global scale, 38% of the earth's surface in 1972 was covered by desert or urban development, versus 17% in 1960.[7]

## Agriculture and Science

*The earth is not a mere fragment of dead history, stratum upon stratum like the leaves of a book, to be studied by geologists and antiquarians chiefly, but living poetry like the leaves of a tree, which precede flowers and fruit—not a fossil earth, but a living earth. . .*[8]
*Henry David Thoreau*

Modern farming owes a great debt to modern science for its high productivity. Yet certain drawbacks are becoming apparent. Agricultural chemists since the middle nineteenth century began to concentrate increasingly on the use of soluble compounds of nitrogen, phosphorus, and potassium as fertilizers. The one sided use of NPK fertilizers depends for "the quantity and quality of the harvest on whether the soil is able to make up whatever else is missing." As proper humus management was neglected, larger amounts of chemicals had to be added to the soil to continue increasing the yields. The scientist W. A. Albrecht found that such unbalanced fertilization damages the quality of fodder, the health of livestock, and increases the danger of plant disease and pests.[9]

The one-sided use of chemical fertilizers as if one were "feeding plants with water-soluble fertilizer salts in a soil which acts like a hydroponic culture" can increase destructive insect life as it destroys the beneficial soil micro-organisms. The loss of humus damages soil structure, erosion grows, and soil fertility begins to suffer. The nitrogen requirement, on soil where nitrate fertilizer is used, grows and upsets the nitrogen balance. In the U.S. corn belt, S. R. Aldrich found that 37% more nitrogen is needed in a corn-soybean rotation than is taken out of the soil by the crop. And the energy required is extremely high. H. H. Koepf reports that:

> In the U.S.A. from 1945-1970 corn yields (America's most important grain crop) have increased from 34 to 81 bushels per acre. The energy requirement for nitrogen rose from 58,800 to 940,800 kcal/acre, and that for all mineral fertilizers together from 74,600 to 1,055,900 kcal/acre. In 1970, 32% of the total energy input for corn production was used for nitrogen and 36% for mineral fertilizer. The amount applied equals about the nitrogen content of the manure of one cow. Spreading 10 tons per acre of manure takes about 398,475 kcal/acre. To produce and spread the same amount of fertilizers (112 lbs. of nitrogen, 31 lbs. of phosphorus, and 60 lbs. of potassium) takes, including spreading, 1,451,425 kcal/acre. An economically significant amount of energy could thus be saved if farm manure were properly utilized.[10]

The modern farmer grows great stands of single crops. These monocultures are valuable because of the ease in harvesting great quantities; yet the very quantity of genetically uniform material is much more susceptible to disease and insects. Greater amounts of pesticides must be applied more frequently. Robert Cahn found that:

> The insecticides caused total destruction of useful insects along with the parasites, and brought about a breakdown of the natural equilibrium within the ecosystem. The pests gradually developed immunity and increased in numbers to a point of producing tremendous damage to crops.[11]

Insecticides have come under increasing attack for the harmful residues they leave, the largely unknown effects of metabolites, and the increasing concentration of toxicity in the food chain. In 1976, Texas Agriculture Commissioner John C. White halted a fire-ant extermination program that used Mirex, a potent insecticide and a suspected carcinogen. The Environmental Protection Agency reports that in areas where Mirex has been heavily used, forty-four percent of the general population already have quantities of this chemical in their bodily tissues.[12]

Plant and animal hybridization also has contributed to increased yields, but this has meant the loss of countless regional and local varieties that had been in use for centuries. Because of their complex hereditary factors these were used for a long time as reserve breeding material. Now we have artificially induced mutations brought about by chemicals or radiation. The living variety of animal breeds and grain, vegetable and fruit strains has given way to a uniformity aimed mainly at high yield.[13]

Through excessive plant-breeding, the great diversity of nature brought into being by various locations, climate and soils has dwindled over the past seventy-five years, as has the chances for further development. In the United States seed societies are forming to collect and grow origin, native seeds. In Russia, a "seed-bank" now collects and preserves such strains of plants.[14]

Chemical farming also seems to damage the nutritional value of food. Dr. Gerhard Schmidt, a Swiss nutritionist, reports that while Switzerland now boasts the highest per capita income among developed nations, its children paradoxically face an anemia rate of thirty-five percent. Schmidt says that while they have a very high protein intake through grains, meat, and dairy products, there is a lack of iron in much Swiss produce such as wheat and barley. Iron must be present in sufficient quantities for the proper use of protein in metabolism. The resulting lack of iron leads to this rather shocking rate of anemia. A major cause is the sharp rise in chemical-based farming, and food processing which reduces the nutritive quality of Swiss produce while increasing the quantity. For example, Schmidt quotes tests that reveal the iron content of biodynamically fertilized plants increased seventy-seven percent in comparison to chemically-fertilized plants. He also states that whole grain bread contains seven times as much iron as white bread.[15]

Carter and Dale wrote that:

> American farmers produced about 40% more food in 1970 than in 1950—on 40 million fewer crop land acres mainly by using better seed and by pouring on the fertilizer. At the same time much of our farmland was eroding and actually getting poorer.[16]

They report that in this twenty-year period, the use of commercial fertilizer rose by 300%.[17] I do not wish to deprecate this tremendous gain in food quantity. Rather, I am asking at what cost socially, ecologically, and nutritionally the gain was made—and at what loss of quality?

### Signs of Social Breakdown

It is often difficult to separate the social and economic aspects of agriculture. A one-sided economic view of agriculture has helped to change the older image of the farmer as the community's producer of food, into a strictly economic one. Thus the results of biological processes—of life processes—must at least partially be handled as any other economic good, as simply so many units of production. Farm labor, animals, plants, and soil life are reduced to the status of inorganic, inanimate goods to be bought and sold. The farmer cannot concentrate on his plants, soils, and animals, but must contend with the vagaries of the marketplace. In agriculture, such thinking has trouble dealing with the myriad variables that are inherent in very complex ecosystems; it works much better in dealing with the more stable inorganic world.

Yet along with the depopulation of rural areas arose a counterbalance in the form of the "back to nature" movement. As the decay and decline of the city becomes more evident, people are moving to suburbs, exurbs, smaller cities, and to the country. For a growing number of people, cities are increasingly undesirable environments in which to live and propagate. In Washington, D.C., abortions now outnumber births. Yet while rural America neither socially nor economically provides a panacea for urban ills, many former city-dwellers flee to the country and adapt to their new home with varying degrees of success.

Other signs of social malaise are the fact that suicide is now the second biggest cause of death between the ages of 15 and 24, and the largest cause on campuses; in 1976, 400,000 young people in this age bracket attempted suicide, twice as many as ten years before. A study done by the Massachusetts Institute of Technology found that the murder rate doubled in per capita frequency between 1964 and 1974. Translated into terms more easily comprehensible, this means that "an American boy born in 1974 is more likely to die by murder than an American soldier in World War II was to die in combat." Further signs of malaise in this age-group were reported in *Time* magazine: in 1975, it reports: ". . . secondary-school students attacked 63,000 teachers, pulled off 270,000 school burglaries, and destroyed school property worth $200 million., In addition, SAT scores for high achievers has dropped since 1962, and overall scores are down 10% for verbal skills and 6% in math. These figures come in spite of a 152% increase in expenditures on education in the past ten years. The murder rates among young people has more than tripled.[18]

### The Common Denominator: The Scientific World-View

I believe that the problems I outlined in agriculture are symptomatic of the failure of the modern scientific world-view to meet these problems. If re-population of rural areas is to succeed, if agriculture is going to turn from its destructive ways, if agriculture is going to provide meaningful employment—then agricultural education is going to have to take a sharp turn, and this change should be predicated on a change in our very *weltanschauung*, our world-conception. Barbara

Ward and René Dubos asked:

> May not the whole development of modern industrial farming represent a dangerous oversimplification; a trend towards a monoculture which, being of its very nature more fragile and vulnerable than balanced, complex ecosystems, exposes mankind to the risk of securing high food returns in the shorter run in return for catastrophic risks of famine later on.[19]

Barry Commoner wrote that

> . . . unlike the automobile, then, the ecosystem cannot be divided into its manageable parts, for its properties reside in the whole, in the connections between its parts. The fault in technology then appears to derive from the fragmental nature of its scientific base. The fault is reductionism, the view that effective understanding of a complex system can be achieved by investigating the properties of its isolated parts.[20]

When we turn our gaze from quantity to quality of life, it appears that we cannot manage problems of life with the present means of knowledge. The scientist and educator Ernst Lehrs described his own realization of this situation:

> On all sides I found this same bewildering gulf between scientific achievement and the way men conducted their own lives and influenced the lives of others. I was forced to the conclusion that human thinking, at least in its modern form, was either powerless to govern human actions, or at least unable to direct them toward right ends. In fact, where scientific thinking had done most to change the practical relations of human life, as in the mechanization of economic production, conditions had arisen which made it more difficult, not less, for men to live in a way worthy of man. At a time when humanity was equipped as never before to investigate the order of the universe, and had achieved triumphs of design in mechanical constructions, human life was falling into ever wilder chaos.[21]

During the past several decades, an increasing number of critics from various branches of our culture have voiced a recurring idea: that our current mode of knowledge—the fruits of which have produced an enormous corpus of information and innovation—is incapable of dealing with *life itself,* and in fact produces a widening rift "between human knowing and human action"[22] Ernst Lehrs said:

> I knew from history that religion and art had once exercised a function which is today reserved for science, for they had given guidance in even the most practical activities of human society. And in doing so they had enhanced the quality of human living, whereas the influence of science has had just the opposite effect. This power of guidance, however,

they had long since lost, and in view of this fact I came to the conclusion that salvation must be looked for in the first place from science. *Here, in the thinking and knowing of man*, was the root of modern troubles; here must come a drastic revision, and here, if possible, a completely new direction must be found.[23] (my emphasis)

If, as I believe, the problems of agriculture and agricultural education are an inherent part of modern life, and if the cause of modern troubles lies "in the thinking and knowing of man," then a discussion of agricultural education must begin here, if it would avoid becoming another learned tome of information but tenuously related to the whole of life. And herein is a chief bias of the writer, that we can achieve an insight into many of life's riddles not merely from the starting point of intellectual history, but the history of ideas seen from the perspective of the evolution of human consciousness. By this I intend to say that human consciousness—the means by which we think and know—is, like we, ourselves, the product of evolution; and we can observe this evolution through the history of our ideas. That one can determine the development of a discipline of knowledge, say that of psychology, through the history of psychological theories, is widely accepted among historians. That through this development one can also demonstrate how human consciousness itself has evolved is a relatively novel concept, yet it has proved to be a fruitful tool of research and it will be shown to be a major tenet upon which a drastic revision of the scientific world-view has been undertaken.

### Goethe and the Evolution of Human Consciousness

In modern intellectual history, the concept of using the history of ideas to illuminate the development of human consciousness originated in the natural scientific work of the poet, dramatist, and novelist Johann Wolfgang von Goethe, specifically in his *Theory of Color*.[24] There, Goethe described the various theories of optics that preceded his own, culminating in Newton's work; Goethe showed how each theory made clear something which had escaped the notice of the previous investigators, and which paved the way for the next step forward. In his writings on natural science, which produced not only his refutation of Newtonian optics but also Goethe's theory of the metamorphosis of plant forms, the great poet was much occupied to show how men could move beyond the limitations which Newtonian science and the philosophies of Kant and Hume placed on human knowledge.[25]

Goethe attempted to demonstrate how succeeding theories built on each other, as man's ability to perceive grew and altered. Modern science, on the other hand, proceeds from the premise that the scope of inquiry into nature continually must he narrowed. As the renowned nuclear physicist Walter Heisenberg wrote:

> Almost every scientific advance is brought at the cost of renunciation, almost every gain in knowledge sacrifices important standpoints and established modes of thought. As facts and knowledge accumulate, the claim of the scientist to an *understanding* of the world in a certain sense diminishes.[26]

Heisenberg decried the fact that this progressive reductionism, while it multiplied the individual facts of knowledge, was itself the product of "renouncing the aim of bringing the phenomena of nature to our thinking in an immediate and living way."[27] Heisenberg felt that Goethe's work made a significant contribution towards a new world-view for science. We will see later that a number of scientific endeavors have emerged from Goethe's efforts.

In his excellent study of Goethean science, *Man or Matter*, Ernst Lehrs found a number of scientists who also felt the need for an enlivened thinking capable of grasping nature in an immediate way. Urging scientists to examine the epistemology of their discipline, the philosopher Alfred North Whitehead wrote that "if science is not to degenerate into a medley of ad hoc hypotheses, it must become philosophical and enter upon a thorough criticism of its own foundations." The physicist Walter Heitler wrote:

> Today when we work in the realm of atomic science or cosmology, or even the most modern realm of physics, the physics of elementary particles, this no longer has the least thing to do with human lie. On the other hand, intellectual thinking is developed to an exceptionally high level . . . If scientific thinking of this character is carried to an extreme and consumes a large amount of human activity, then it is perhaps understandable that it often, although not always, must be at the expense of other things, for example the realm of feeling. *Here one can speak of a certain kind of destruction of the human soul.*[28] (Heitler's emphasis)

Sir Arthur Eddington said that "The sciences of inert matter have led us into a country that is not ours . . . Man is a stranger in the world he has created." The historian Leo Marx wrote that "The current environmental crisis has in a sense put a literal, factual, often quantifiable base under this poetic idea "—the need for human harmony with nature, a need often stressed in the past by poets and now called for by scientists as well.[29]

As we read daily of mercury poisoning, PBBs, PCBs, DDT, Mirex, radioactive waste, spills and fall-out, and the growing number of other serious pollutions of the fragile biosphere of our planet, do we not find ourselves increasingly alienated from our environment both physically and psychologically?

So it was Goethe who began the methods of perception and cognition with which modern man can escape the limitations proscribed by Hume, Kant and Newtonian physics, which have led to what René Dubos called "the dehumanization of the scientist."[30] In doing so, he also began the study of the development of consciousness itself because he wanted to show that the human being was a product of evolution, that his present condition was only one stage in a process of becoming. Kant, for example, held that man's cognitive capabilities at that time were his permanent state of being. Clearly then, Goethe, the discoverer of the principle of metamorphosis in Nature, could not accept stasis as a given human quality. As Ernst Lehrs put it:

> Goethe started from the conviction that our senses as well as our intellect are gifts of nature, and that, if at any given moment they prove incapable through their collaboration of solving a riddle of nature, we must

ask her to help us develop this collaboration adequately.[31]

Rather than progressively restrict our senses to the limits of our intellect, Goethe urged us to expand our intellect to new levels of cognition. He said: "The senses do not deceive, but the judgment deceives." Lehrs states that:

> Goethe's path was aimed at awakening faculties, both perceptual and conceptual, which lay dormant in himself. His experience showed him that "every process in nature, rightly observed, awakens in us a new organ of cognition." Right observation, in this respect, consisted in a form of contemplating nature, which he called a "re-creating (creating in the wake) of an ever-creative nature."[32]

But Goethe never gathered his natural scientific writings into a cohesive philosophy. It fell to Rudolf Steiner to provide the philosophical theory of knowledge for Goethe s ground-breaking work. Steiner built his own philosophy, which he called "Anthroposophy" on his continuation and enhancement of Goethe's methods of cognition.

**Footnotes for Chapter One**

1. For the failure to compute the social costs of technological development, see Barry Commoner, *The Closing Circle: Nature, Man, and Technology* (New York: Alfred A. Knopf, 1972) 21-27.
2. Henry A. Wallace, quoted in Hartmut von Jeetze, "Biodynamic Relations Between Man and the Land," *Biodynamics*, 115 (Summer 1975) 11.
3. Ibid. 12.
4. Mike Clary, "The Dilemma of Michigan's Family Farms," *Parade* (October 27, 1976) 3. Frans Carlgren, *Education Towards Freedom* (East Grinstead, Sussex: Lanthorn Press, 1973) 107.
5. Clary, Ibid.; Dan Maguire, "A Critical View of Techno-industrial Agriculture," *Biodynamics*, 107(Summer 1973) 25.
6. Commoner, *The Closing Circle*, 81-82.
7 Vernon Gill Carter and Tom Dale, *Topsoil and Civilization*, rev. ed.(Norman: University of Oklahoma Press, 1974) 59; Lloyd Shearer, "Agricultural Revolution," *Parade* (October 27, 1976) 10.
8. H.D. Thoreau quoted in The Chambers Family, "The Horns of a Dilemma," *Biodynamics*, 108(Fall 1973) 4.
9. Koepf, Pettersson, and Schaumann, *Biodynamic Agriculture: An Introduction* (New York: 1976) 3-4.
10. Maguire, "A Critical View of Techno-industrial Agriculture," 15; Koepf, Pettersson, and Schaumann, Ibid., 159-160.
11. Robert Cahn, "The Pitfalls in Aid Projects," *Christian Science Monitor* (19 December 1968) quoted in Maguire, Ibid.p 28.
12. Koepf, Pettersson, and Schaumann, *Biodynamic Agriculture*, 5-6; Bob Calvary, "Man Now Bringing Plagues on Himself," *Detroit Free Press* (14 November 1976) F4.
13. Koepf, Pettersson, and Schaumann, Ibid., 4.
14. B. Likhachev, "A National Seed Bank," *Biodynamics*, 121 (Winter 1977) 31-32.
15. Gerhardt Schmidt, "Aspects of Protein Nutrition," *Biodynamics*, 130(Fall 1976) 1-6; Schmidt, "Nutrition and Agriculture," lecture given at the Alternative Agriculture Conference, Wilton, New Hampshire, I August 1976. See also Ross Hume Hall, *Food for Thought: The Decline in Nutrition* (New York: Vintage Books, 1976) for an excellent discussion of the problems in nutritional science. Hall traces reductionistic, materialistic thinking in nutritional science and in

agriculture. See also Maria C. Linder, "A Review for the Evidence of Food Quality," *Biodynamics*, 107(Summer 1973) 1-12.
16. Carter and Dale, *Topsoil and Civilization*, 234.
17. Ibid., 237.
18. CBS News "Special Report," (31 November 1977); "War Today?", *New Times* (30 April 1977) 37; and "High Schools Under Fire," *Time* (14 November 1977) 63, 62.
19. Ward and Dubos quoted in Koepf and Budd, "Biodynamic Agriculture," in John Davy, ed. *Work Arising from the Life of Rudolf Steiner* (London: 1975) 166-167.
20. Commoner, *The Closing Circle*, 187-188.
21. Ernst Lehrs, *Man or Matter*, 2nd ed. rev. (London:Faber and Faber, 1958) 24.
22. Ernst Lehrs' phrase. Some excellent studies of this problem are: René Dubos, *The Dreams of Reason: Science and Utopias* (New York: Columbia University Press, 1961) and his *So Human an Animal* (New York: Charles Scribners' Sons, 1969); Sir Arthur Eddington's several works, such as *New Pathways in Science* (London: Faber and Faber, 1949) and *The Philosophy of Physical Science* (Ann Arbor: University of Michigan Press, 1958); and Alfred North Whitehead, *Science and the Modern World* (New York: Harper and Row, 1942).
23. Lehrs, *Man or Matter*, 25.
24. J.W. von Goethe, *Theory of Colours* (Cambridge: Cambridge, University Press, 1970).
25. For this discussion of Goethe's contribution to the history of ideas I am indebted to a lecture by Dr. Hagen Biesantz, "Changing Consciousness Through The Ages," given 6 November 1976 at the University of Michigan. See J.W. von Goethe, *The Metamorphosis of Plants* (Springfield, Ill.: Biodynamic Farming and Gardening Assn., 1974).
26. Walter Heisenberg quoted in Lehrs, *Man or Matter*, 33.
27. Lehrs, Ibid., 34.
28. Walter Heitler quoted in Carlgren, *Education Towards Freedom*, 132.
29. Arthur Eddington cited in Lehrs, *Man or Matter*, 35; Leo Marx, "American Institutions and Ecological Ideals," *Science*, 170(September 1970) 947.
30. René Dubos, *The Dreams of Reason*, 129.
31. Lehrs, *Man or Matter*, 84.
32. Ibid.

## CHAPTER TWO: RUDOLF STEINERS LIFE AND WORK

### The Scope of Steiner's Work

Although Rudolf Steiner's work built a world-conception out of which a growing number of people in diverse cultural activities are taking steps toward a renewal of our culture, the very magnitude of scope and quantity of his output make an evaluation difficult, especially so in our era of specialization. Steiner, who lived from 1861 to 1925, wrote twenty-three books and delivered over 6,000 original lectures; the majority of these lectures had an educational character and exerted a formative influence on a variety of cultural institutions. In his *The Dignity of Man*, Russell Davenport, for many years the editor of *Fortune* magazine, integrated Steiner's ideas with those of major American thinkers. He wrote:

> That the academic world has managed to dismiss Steiner as inconsequential and irrelevant is one of the intellectual wonders of the 20th century. Anyone who is willing to study these vast works (let us say a hundred titles) will find himself faced with one of the greatest thinkers of all time, whose grasp of the modern sciences is equaled only by his profound learning in the ancient ones. Steiner was no more a mystic than Albert Einstein; he was a scientist rather—but a scientist who dared to enter into the mysteries of life[1]

Today, over twenty years alter Davenport voiced his complaint, the situation is somewhat altered. The University of Michigan is one major center of learning which has regularly offered course work on Steiner's cultural contributions. Antioch College of New England has a Master of Arts program based on Steiner's pedagogy. The Waldorf Institute in Spring Valley, New York offers two professional teacher training programs (early childhood and primary through high school, with a Master of Arts degree) and a thirty-eight semester hour interdisciplinary year devoted to Steiner's contributions to fields other than education, reviewed in light of current research. A course in Steiner's work in education for the handicapped will begin in September, 1991. Emerson College, in Sussex, England, offers several year-long programs based on Steiner's work, including biodynamic agriculture, education and social development. Rudolf Steiner College in Fair Oaks, California, offers courses in the arts and education and a foundation year in anthroposophical studies. Professional courses and adult education programs are available in many other countries. Medical clinics, businesses such as the Weleda pharmaceutical firm, and counseling services are available in many places.

Today, then, we find a growing recognition of Steiner's contributions to Western culture, on which were founded the numerous homes and villages that provide help according to Steiner's indications for mentally retarded or emotionally damaged children; the Waldorf School movement, which now has over one hundred schools in twenty nations, and the curriculum which the one hundred and five state schools of the Canton of Bern, Switzerland, have adopted; the several hundred farms that employ the biodynamic methods which Steiner formulated; and the unique building called the Goetheanum, which he designed near Basel,

Switzerland. In Europe several teaching and research centers conduct extensive scientific and artistic work. These include the Lucas Clinic, Ita Wegman Clinic, and the Weleda Pharmaceutical Company, Arlesheim, Switzerland; the Eurythmeum, Stuttgart, Germany; the Friedrich Husemann Clinic, Freiborg, Germany; the Park Attwood Clinic, England; the Rudolf Steiner Seminariet, Jarna, Sweden; and the Goetheanum, Dornach, Switzerland, which is the international center for all these activities, and which offers courses in eurythmy, the plastic arts, painting, agriculture, and other studies. In a recent book on Anthroposophy, the cultural historian Stewart C. Easton[2] wrote:

> If Steiner had been nothing but a philosopher, or theologian, or educator, or authority on Goethe, or agricultural expert, or architect, or knowledgeable in medicinal plants, or dramatist, or gifted artistic innovator, inventor of eurythmy, an age that respects specialization would have reserved a special niche for him. But Steiner was all these things at the same time.[3]

**Steiner's Early Years**

Rudolf Steiner was born in Kraljevec, Austria, on February 27, 1861, where his father was the railroad station-master, and his early years were spent in villages along the railroad.[4] By the age of eight, Steiner had learned to distinguish between things which are "seen" and those which are "not seen." "For the reality of the spiritual world was to me as certain as that of the physical. I felt the need, however, for a sort of justification of this assumption," Steiner wrote, and he found through geometry "the first budding forth of a conception which later gradually evolved within me."[5] This conception, which lived unconsciously in Steiner at this time, took "a definite and fully conscious form" in about his twentieth year:

> The objects and occurrences which the senses perceive are in space. But, just as this space is outside man, so there exists within man a sort of soul space which is the scene of action of spiritual beings and occurrences. I could not look upon thoughts as something like images which the human being forms of things; on the contrary, I saw in them revelations of a spiritual world on this field of action of the soul.[6]

Steiner's concern with finding the "unseen" spiritual world within the "seen" physical world was "a burning question which became a dominant theme of his first thirty years."[7]

When he was about ten years old, a teacher introduced Steiner to the basic principles of Copernican astronomy, which he added to his growing conception of the world. At the age of eleven, he entered the Realschule at Wiener Neustadt, where he studied the sciences. Steiner became convinced that he must:

> grapple with nature in order to acquire a point of view with regard to the world of spirit which confronted me in *self-evident perception*. I said to myself that it is possible after all to come to an understanding of the experience of the spiritual world through one's soul only if one's process of

thinking itself has reached such a form that it can attain to the reality of being which is in the phenomena of nature.[8] (My emphasis)

During the next few years, Steiner learned, partly on his own, a considerable amount of analytical geometry and probability calculus, and tutored fellow pupils to help his family's financial situation. Because the school prepared its students for the sciences, Steiner received no knowledge of classical languages; so he bought Latin and Greek texts which he mastered sufficiently to be able later to coach other students. He also read deeply in church dogma, symbolism, and ecclesiastical history. He graduated from the Realschule with honors at the age of eighteen.[9]

During these years, Steiner felt even more strongly that: thinking can be developed to a faculty which actually lays hold upon the things and occurrences of the world. "A 'substance' which remains outside of thinking, which we can merely reflect about, was to me an unendurable conception. Whatever is in things, this must enter into human thought. . ."[10] This statement sounds remarkably like Goethe's description of his scientific methods. And as Goethe had confronted Kant's philosophy many years before, so Steiner, although he was still an adolescent, worked his way through Kant's *Critique of Pure Reason*. It should be clear by now how Steiner's own experiences made it impossible for him to accept Kant's epistemological limitations. In the next period of his life, Steiner came to Goethe's scientific work and began the formulation o! his own theory of knowledge, based on the methods of modern science and on the growing powers of clairvoyance which had begun early in his life.

### College Years and Goethe

Steiner entered the Vienna Polytechnic, then one of the most important scientific institutions in the world, where he studied biology, chemistry, physics, and mathematics. He also attended many lectures at the University of Vienna, including those by the philosopher Franz Brentano. His own philosophical studies moved from Kant to Fichte's *The Science of Knowledge*. Fichte's philosophy of the ego, which had influenced other thinkers such as Nietzsche and Kierkegaard, helped Steiner to mold into the language of concepts the direct, immediate perception of the spiritual world which he possessed. Now he felt the need to take this direct perception of the Ego as a spiritual being existing in a spiritual world, into the world of nature. Unlike most modern men, he had ready access to the world of the spirit; but he struggled to discover how to bring these perceptions into the world of nature.

Two important encounters at this time helped to give Steiner the proper direction in his quest. He met an old peasant herb-gatherer, Felix Kogutski, who was without formal education, but whose faculties of spiritual perception enabled him to see deeply into the secrets of nature and to know the curative properties of the plants he gathered and sold to Viennese doctors. Steiner felt that the old man was "a soul from ancient times," a last representative of "an instinctive clairvoyance of an earlier era," and he relished the herb-gatherer's intuitive knowledge of the spiritual realities of nature. This man was the first person with whom Steiner felt able to discuss his own spiritual life.[11]

The other meeting was with the famous Goethe scholar, Karl Julius Schroer, who lectured regularly at the Polytechnic, and in whom Steiner met a vigorous advocate of idealistic philosophy. Schroer's influence led Steiner to deepen his study of Goethe, and his work impressed Schroer sufficiently that he recommended his pupil to the publisher Joseph Kürschner, who was leading a staff of scholars in preparing a complete edition of Goethe's works. Then only twenty-one, Steiner was by far the youngest of these editors, yet his introductory essays to Goethe's scientific writings are models of clarity and exhibit an astounding breadth of knowledge. These volumes appeared in 1984, 1887, and 1890.[12]

Steiner gathered his thoughts on Goethean science into his first book, *Outline of a Theory of Knowledge Implicit in Goethe's World-Conception* (1886), in which he supplied the epistemological basis for Goethe's scientific method; he later expanded these ideas in his *Goethe's Conception of the World* (1897) one of Steiner's major philosophical works. Through the efforts Karl Julius Schroer, Steiner was appointed editor at the Goethe-Schiller Archives in Weimar, Germany, where he worked from 1890 to 1896. During this time, Steiner edited the remaining Kürschner edition volume on Goethe, seven volumes of another Goethe edition, edited the works of Schopenhauer, Jean Paul Richter, Wieland, and Uhland. He wrote books on Nietzsche and the scientist Ernst Haeckel, a doctoral dissertation, and his own epistemology, *The Philosophy of Spiritual Activity*.[13]

## Steiner's Theory of Knowledge

Steiner's doctoral dissertation for the University of Rostock dealt with Fichte's epistemology, and was published as *Truth and Science* in 1891. In this work, Steiner attempted "the reconciliation of the philosophical consciousness with itself", by which he meant that a true theory of knowledge must begin by investigating the basis for all philosophical activity - the activity of thinking itself. Steiner's own highly innovative epistemology, *The Philosophy of Spiritual Activity* (1896), contains in purely philosophical form the epistemological basis for all his later work. This book sought to develop a science o mental activity by the natural scientific methods of rigorous thinking and careful observation. In Steiner's life *The Philosophy of Spiritual Activity* marks a maturation of his own spiritual striving and of his philosophy.[14]

Before turning to the remainder of Steiner's life and its bearing on this dissertation, I must discuss the manner in which he built on Goethe's work to develop the philosophical basis for a renewal of science and the arts. Steiner sought to provide an all-inclusive world-view, or world-conception, which would answer two fundamental questions of human existence. As he wrote in his preface:

> Is it possible to find a view of the essential nature of man such as will give us a foundation for everything else that comes to meet us—whether through life experience or through science—which we feel is otherwise not self-supporting and therefore liable to be driven by doubt and criticism into the realm of uncertainty? The other question is this: Is man entitled to claim for himself freedom of will, or is freedom a mere illusion begotten of his inability to recognize the threads of necessity on which his will, like any natural event, depends?[15]

If mankind could answer these questions affirmatively, Steiner thought, then a new bridge could be built between the separate sciences, the arts, philosophy, and religion. Steiner wanted to meet squarely the problems of knowledge as I have outlined them in this essay. The scientific thinking that had arisen threatened then, as it does now, to become the only means of knowing which men recognized as producing "valid" knowledge. An a scientist, Steiner was quick to applaud the gains in knowledge which science provided for humanity; yet the intellectual, analytical thinking with which scientists pulled apart and learned the secrets of the inorganic world, seemed unsuited for understanding the organic, living world. Scientists can understand how a living organism transforms matter into energy only by analyzing the various individual processes into their chemical constituents, and this analysis can proceed down to the most minute particles, until we lose sight not only of the original organism's living nature, but even of the nature of the tiny particles. As Sir Arthur Eddington said:

> This spectacle is so fascinating that we have perhaps forgotten that there was a time when we wanted to be told what an electron is. The question was never answered. No familiar conceptions can be woven round the electron: it belongs to the waiting list. The only thing we can say about the electron, it we are not to deceive ourselves is: "Something unknown is doing we don't know what."[16] (His emphasis)

Perhaps this uncertainty is allowable in the natural sciences, but when it becomes the criteria of knowledge !or all human endeavor, the results stimulate the chaos we see everywhere in our civilization.

Steiner was convinced that man could step outside the boundaries of knowledge if knowledge itself could become organically alive. Since life is itself a unity, "the more deeply the separate sciences try to penetrate into their separate realms, the more they withdraw themselves from the vision of the world as a whole."[17] Steiner asserted that the sciences are valuable stages toward building a new epistemology. He compared the scientific philosopher to a composer, who follows the rules of musical composition, in which the rules of the theory "become the servants of life itself, of reality." All real philosophers, he said, have been "artists in the realm of concepts."[18] Where Goethe had applied clear thought, accurate observation, and intense meditation to the realm of nature, Steiner turned these techniques inward onto thinking itself.

Rudolf Steiner began his study with a survey of human action, as seen by such philosophers as Spinoza and von Hartmann. He held that to have a concept of an action for which the reason is known presupposes thinking, without which it is impossible to formulate concepts of knowledge about anything. Since thinking produces the mental pictures that form the motives for actions that rise above the mere satisfaction of animal desires, Steiner stated that a discussion of the origin and nature of thinking must precede an investigation into action.

Steiner thought that dualistic philosophy arose from the barrier that we erect as soon as we reach a certain level of consciousness between our Ego, or "I", and the object of perception. The feeling that we are still part of the universe remains, however, and our attempt to bridge this chasm has resulted in the "whole spiritual striving of mankind".[19] The dualist considers only this separation, while the monist attempts to discount it. To bridge the gulf, Steiner studied what aspects of the world—of nature—we have brought with us in our flight from it.

He found two activities to be present in all conscious spiritual striving: observation and thinking. Observation precedes thinking; in fact it is through the observation of ourself as "subject" and the world as "object" that the question of separation arises in the first place. But thinking usually is not observed, and can only be observed by what Steiner called "an exceptional state of consciousness," because our own activity produces it. What we usually see is not thinking itself, but the mental picture which thinking produces. Since we create thinking, said Steiner, it is not "given" like other objects in the world; thus it is the most accessible and immediate activity of our being. Steiner therefore maintained that thinking must be our point of departure for studying our percepts of the world. In other words, "I" am an object that exists in a sense I can derive from myself, since I give to my existence the self-determining, self-supportive content of my thinking activity.

Steiner wrote that our consciousness is the mediator between thinking and observation. Consciousness itself, awareness, is passive; thinking is an active process that divides the world into subject and object. When we focus our thinking on observation, objects arise in our consciousness. Thinking directed toward the self gives rise to self-consciousness— the concept of "I" as subject. Here Steiner showed again that the ego, as thinker, erects the barrier between the self and the world.

Thinking goes beyond mere observation by linking observations together through concepts, but Steiner emphasized that concepts and observations come from two different directions. When a concept arises to meet a percept, a new mental picture is formed, but many philosophers fail to realize that a mental picture is a percept, too: it is known through observation. According to Steiner, most philosophers mistake the seeing of a mental picture for thinking itself, when in actuality the mental picture is an abstraction, or residue, from the active thinking process. Thus, following Kant, many philosophers maintain that thinking can produce no knowledge of the "thing-in-itself," but only knowledge of the thinker's mental pictures.

Steiner pointed out that we experience directly our percepts, including mental pictures and our feelings, and that we feel they belong together; yet our first experience of thinking is the mental picture left after thinking has taken place. Rudolf Steiner held that actually a mental picture cannot come between thinking and the thinker; cognition transcends both subjective percepts (mental pictures) and objective, visible percepts (trees, cars, dogs). Thinking, he said, is not limited like man and his perceptions, but is universal and all-pervading. Man creates his concepts by his thinking as little as he creates objects by perceiving them; rather, through thinking, concepts arise out of the "divine world-continuum" of ideas, and thinking brings these ideas into our consciousness, where they appear as mental pictures.

For Steiner, then, thinking is the synthesis of the percept with a concept that arises to meet it. Thinking also combines separate percepts with concepts into a coherent whole; if it were not for this, our world of percepts would be a whirling series o! unconnected thoughts and sense-perceptions. Steiner emphasized that thinking is never apart from the object, but is an organic part o! the world process: while thinking creates the separation of "I" and the world that leads to the fundamental desire for knowledge, thinking through its universal character heals this separation. It allows man to ascend by his own activity to the world of ideas, a world beyond subjectivity. In this way cognition itself becomes a mode of perception.

Steiner asserted that human individuality occurs in two ways: through our mental pictures which are individualized images of concepts, formed by thinking; and by feelings:

> Thinking is the element through which we take part in the universal cosmic process; feeling is that through which we can withdraw ourselves into the narrow confines of our own being.[20]

A one-sided feeling life devoid of thinking leads to solipsism. Steiner urged mankind to reach with their individual feeling life as far as possible into the realm of ideal concepts. He wrote that "feeling is the means whereby, in the first instance, concepts gain concrete life."[21] Ideal feelings make cosmic thought into human thought; they breathe life into pure ideas.

We have now before us in the barest outline, the picture Steiner gave of the means by which humanity can obtain knowledge, both of the physical world and the world of the spirit. Only man's perceptual organization, he stated, limits his acquisition of knowledge. Once he sharpens his observation sufficiently to achieve the "extraordinary state" by which he can observe concepts as they appear in his consciousness, then the human being can observe in a new way. He can observe with a quiet mind, so that the perception can speak and so that the proper concept will emerge to meet it. For humanity is the mediator, the ground of consciousness on which the perceptual element of objects merges with the conceptual element. We extract with our thinking the conceptual element, and in our thinking "re-create" reality by merging the concept with the percept.

## An Enlivened Thinking

In time, with proper training, a warm, living, holistic thinking can emerge which reveals the reality behind sense perceptions and mental pictures. For example, through thousands of observations of the living world, Goethe entered deeply into the life of plants and penetrated with his cognition and observation into the way in which they changed their forms. He discovered the laws that governed plant growth—"living" ideas that he could see with the "eyes" of his imagination. In the plant world Goethe "saw" the archetypal plant which existed nowhere in nature but from which all plants are in some measure derived. This idea allowed Goethe to investigate how the archetype expressed itself in so many different types of plants, so apparently different to the untrained sense perception.[22] Goethe's theory of plant metamorphosis is generally accepted today, yet how different this method of knowledge appears from the abstract, theoretical thinking of most scientists, how different from the usual "laws" of the physical-material world, derived by analyzing bits of plant tissue torn from the living plant.

Rudolf Steiner called this living, organic thinking "Imagination", and he stated that the majority of people in our age who try to develop this cognitive ability will be able to do so. In his most widely read book, Steiner wrote that "there slumbers in every human being faculties by means of which he can acquire for himself a knowledge of higher worlds."[23] Humanity's present powers of cognition and perception have evolved over time; through diligent efforts, humanity can speed the process of evolution and increase these abilities. Beyond imaginative thinking, Steiner spoke of two higher realms of cognition, which he called Inspiration and Intuition. These faculties he developed in himself and this enabled him to teach the results of his super-sensible research in a form accessible to human thinking and human understanding. While many people can develop imaginative thinking, few people have the ready, constant and fully conscious access to inspirations and intuitions that Rudolf Steiner possessed; yet he stressed that all men have the ability to ponder the teachings from the higher worlds of cognition and he held that this is an important step toward awakening faculties which can lead to an individual perception of spiritual realities.[24]

### The Foundations of Anthroposophy

Now we turn again to the thread of Steiner's biography. He left Weimar in 1897 and moved to Berlin, where he edited an avant-garde literary journal and was in increasing demand as a lecturer. For seven years he taught at the Workers' College, because he deeply felt the lack of culture of the working class people. These people had within one or two previous generations lived in a rural peasant culture the roots of which extended hundreds of years into the past, and which possessed the deepest spiritual roots. Steiner taught them history and public speaking, so that they could gain an appreciation for their place in history and develop the ability to express themselves in clear, well articulated thoughts, thereby regaining a measure of autonomy. These efforts produced many life long admirers for Rudolf Steiner, but he was eventually asked to leave the college when it became apparent that his lectures, based on the idea that each human being is a spiritual being, and that there are spiritual currents running beneath the outer aspects of history, would never conform to orthodox Marxist materialism.

After the turn of the century, Steiner began writing the basic books of his own world-conception. These works include *Mysticism at the Dawn of the Modern Age and its Relation to Modern World Philosophy*, *Christianity as Mystical Fact and the Mysteries of Antiquity*, *Theosophy: An Introduction to Supersensible Knowledge in the World*, *Knowledge of Higher Worlds and Its Attainment*, and finally *Occult Science—An Outline*.[25] In a certain sense this phase of Steiner's life ends with the publication in 1914 of his monumental historical study *The Riddles of Philosophy*, perhaps his crowning achievement as a philosopher. Building on his epistemology and his own ever-expanding perceptions, Steiner sought to unify the divergent cultural streams of our century. His lecture series included cosmology, history, philosophy, education, theology, and mathematics.[26] As one long time student of Steiner's thought characterized these efforts:

> (they were) to rear an edifice of spiritual knowledge equal in accuracy to, but wider in scope than, natural science, a knowledge based not on intel-

lect alone but on the full gamut of human faculties. This edifice embraced universe, earth, and man. It contained, indeed, not a little ancient knowledge jettisoned by the science of the modern age, but given fresh form and significance. It is therefore fundamentally a new knowledge, new in conception, new in presentation, new in the demands it makes upon those who seek to understand it. It is a knowledge which, as its name, Anthroposophy implies, regards man not as a peripheral accident, but as central to the understanding of the entire universe.[27]

The seven years between 1910 and 1917 saw Steiner devote increasing attention to the arts. He wrote, produced, and directed four original dramas in which he depicted the spiritual life of modern times and patterns of the reincarnations of individuals. These dramas are regularly performed today in Europe and America. Steiner designed and supervised the construction of the first Goetheanum, a theater building built in a unique style of organic architecture.[28] In 1912, he invented the art of movement he called Eurythmy, which today is widely used as a stage art, as an educational aid in schools, and as a therapeutic method.[29]

Between 1917 and the beginning of his last illness in September, 1924, Rudolf Steiner gave a number of lecture cycles to professional groups which began the various activities I mentioned at the beginning of this biographical sketch. These groups included doctors, educators and teachers, clergymen and theologians, actors and stage directors, those people engaged in remedial education, and agriculturalists. In 1917, Steiner presented what a number of people consider his most important scientific discovery, a conception of the human being which he called the "threefold human organism," and which he had researched for over thirty years previously.[30] I will provide a brief exegesis of this theory since Steiner based much of his pedagogy, medical work, and agricultural work on it. The concept of the threefold organism can serve as a first introduction to his work in education and an example of Steiner's thinking in a different direction from *The Philosophy of Freedom*. There he attempted to build a foundation for the psyche to examine its own activity, pure thinking. Here he tried to construct a bridge from the psychological to the physiological being of humanity.

## Steiner's Threefold Conception of the Human Being

Rudolf Steiner considered the human being as a totality, and he held that "the whole organism was an organ o! consciousness. However, within the unity of this totality, he discerned a polarity of functional systems. He also described a third system acting as a mediator between the two."[31] Steiner related the free movement of the limbs to metabolic processes. The limbs themselves are organs of catabolism (destructive metabolism) and the digestive organs are the center of anabolism. This "metabolic-limb system" was for Steiner a functional interaction which included all the oxidations of substance which occur in muscle movement, neural functions, and indeed anywhere in the organism.

Steiner asserted that this system served as the physical foundation for "the least conscious element in the human psyche", the will.[32] Not surprisingly, the processes of metabolism occur outside the realm of consciousness unless there is some dysfunction, such as indigestion; then the metabolic process becomes painfully conscious. Many routine activities are of a "will" nature, such as driving a well-known route or brushing one's teeth. We are often conscious of quite other things while performing these tasks.

For Steiner, then, "the will formed the polar opposite of conscious thinking in every respect."[33] The physical foundation of thinking Steiner held to be the non-organic substances of the nervous system, which accords with modern medical theory. Unlike the metabolic-limbs system, the nervous system has no moving parts; instead we see a minimum of movement and life. The nerves and brain undergo a constant process of destruction. Harwood says that "it is this death process, this suppression of life, which enables individual consciousness to penetrate and find its support in the physical body."[34] Steiner maintained that as the nerves carry ideation and consciousness to all parts of the body, so the blood carries the will. The two systems interpenetrate each other in all parts of the body through the interaction of the nerves and the blood system. Here we car see the wisdom in older expressions such as "blinded by rage" in which anger causes an oversupply of blood to the nervous system, resulting in a temporary dimming of consciousness. This example brings us to the mediator between the two polar systems of metabolic-limb and nerve-sense: the rhythmic system, which includes breathing, circulation, and the endocrine glands. Steiner held that this system supplies the bodily basis for feeling in the psychological realm. In terms of consciousness, feeling occupies a middle area between the unconscious will and conscious thinking; Steiner compared feeling to dream. "Midway between the crystalline clarity of thought and the activity of the will," feeling enlivens thought and draws the deeds of will into consciousness.[35]

Before I describe how Steiner used this psychosomatic concept of humanity in education, perhaps it is well to note that he in no way disparaged the knowledge gained through physiological research. Steiner's idea of the evolution of consciousness was such that he thought the separation of psyche and soma was a necessary step to enable humanity to come to a conscious understanding of the relationship between them. For Steiner, the same stage in the evolution of consciousness that has produced the glorious discoveries of the sciences also produced the many "separations" which I mentioned earlier. The threefold concept of the human being attempts to unify psychology and physiology, just as Steiner's earlier work tried to re-unite philosophy and science.[36]

## Steiner's Work in Education

In education, Rudolf Steiner spoke of three rhythmically successive periods of maturation which he related to the threefold psychosomatic model. These periods last about seven years in length. The first is the time from birth to the change of teeth, when the child lives totally in movement—even his thinking is concerned with what things do, or what you do with them.[37] Steiner calls this the unfolding of the will activity, and the somatic correlative of this is the formation of his internal organs. The child learns almost totally through *imitation* in this first phase of development, by imitating the activity and sounds of his environment. Steiner urged educators to help children build their bodies to support a strong and healthy will by concentrating on activities the child can imitate and in which rhythmical activity is emphasized.

At about the time of the change of teeth, usually around the seventh year, a new phase begins. The body is formed, and forces which were used almost exclusively for that task are freed to support increased conscious activity. This seven year period is the time for learning through feeling. The emphasis is on presenting the lessons in an artistic way, with much care given to providing a rich image content through stories, myths and legends; Steiner held that this was the age of *imagination.* While intellectual subjects are offered, such as arithmetic, geography, or history, the teacher attempts to teach in as pictorial a manner as possible. In a history lesson, for example, the class would hear the story of Caesar crossing the Rubicon, then draw it for their lesson books or paint it, and then perhaps act it out. Steiner maintained that feeling affects the total child—the conscious thinking, semi-conscious feeling and unconscious will. A thing known only by the head is only partly known, and time is required before real feeling takes root in the child. Thus the teacher appeals to the thinking through a lively, imaginative presentation of the subject and then engages the limbs through drawing, writing, or modeling. In this manner the teacher brings about a movement from head to heart, and from heart to hands.[39]

The next seven-year period usually begins about the time of puberty. Here for the first time, according to Steiner, sufficient forces are present in the psyche for real intellectual work. Now the child can begin to tackle the fruits of modern intellectual consciousness. The progressive liberation that began at the change of teeth can now be used for thinking; not the unconscious will activity of early childhood, nor the dreamy, imaginative feeling-life of the second period, but the intellectual thinking which Steiner held was the province of our waking consciousness.

I will conclude this brief rendition of Steiner's pedagogical ideas by emphasizing their evolutionary, holistic aspects. The reader can see how Steiner felt that a proper pedagogy must be fitted to the child's total psychosomatic development, not merely to a premature stimulation of his intellect. Indeed, Steiner argued that such an early over-emphasis on the child's thinking could not only hamper the development of his will and feeling, but could very well lead to organic damage as well. The story curriculum also exhibits this developmental character. In kindergarten and first grade, fairy tales are told; in second grade, folk tales and fables are used; Old Testament stories are added in the third grade; the fourth grade moves on to Norse mythology; the fifth grade is told Greek myths; and the sixth grade takes up Roman history. For Steiner, the child should follow the development bf the human race, which has evolved from fairy tale, to fable, to mythology, and finally to history.[40]

## Biodynamic Agriculture

In 1924, Rudolf Steiner gave the lecture series which inaugurated the biodynamic farming movement, although his assistants had begun experimental work three years earlier.[41] In these lectures, Steiner discussed the nature of the earth as a living organism, the rhythms of the cosmic and terrestrial forces and elements, and he gave essential descriptions of the most important substances in soil cultivation and human nutrition. Steiner related his threefold model of the human organism to the plant and to the animal, an aspect of his teaching which has had far-reaching consequences for both agriculture and for medicine, and which I will discuss in some detail in a later section of this book. He provided specific, concrete suggestions for managing a farm as a living organism, as a totality. No special training in Steiner's philosophy is necessary for the farmers who apply these methods, but their use

> gradually leads the one who uses (them) to another world-picture, that he to begin with learns particularly to judge the biological (i.e. life) processes and interconnections in a different way than the materialistic chemically inclined farmer. . . he will bring to the dynamic play of forces in nature a greater degree of interest and awareness. . . There is a difference between mere application of the method and *creative collaboration*.[42] (My emphasis)

Biodynamic agriculture involves administering the earthly and cosmic forces to assure healthy, nutritious soils and plants. By earthly forces Steiner meant the chemical substances in the soil and the atmosphere which nourish plants and provide the physical substance or growth. By cosmic forces he meant the light of sun and moon, the rhythms of sun, moon, and planets, and the atmospheric warmth.[43] The farmers make two groups organic preparations which facilitate the activity of these polar opposite forces. Six of the dynamically effective substances are added to compost and other farm manures, which the farmer uses in lieu of chemical fertilizers. These preparations greatly increase the number of nitrogen-fixing bacteria; increase the amount of trace elements which in turn stimulate bacteria growth; enhance root growth and humus-forming processes in the earth; and stimulate stem and leaf growth. One preparation is sprayed directly onto the soil to stimulate soil bacteria life; another substance, made with powdered quartz, is sprayed onto the green leaves to enhance their ability to utilize sunlight.[44]

Biodynamic farming methods are directed at increasing the life in the soil. The farmer views his farm as a living organism which must be managed in a living way. He does not resort to poisons or chemical fertilizers which have deadly results for soil life, destroying the beneficial organisms on which humus formation depends, and which cause the soil to become progressively mineralized. This mineralization in turn demands even higher concentrations of pesticide and fertilizer which necessitate larger expenditures for equipment and chemicals, until the vicious cycle can terminate in bankruptcy or the farmer, lifeless soil which erodes easily by wind or rain, dire consequences for nutritive produce, and water and soil pollution.[45] An extensive literature of experimentation and practice has grown in the past fifty years out of the indications which Rudolf Steiner gave in 1924. A number of educational endeavors have arisen which employ Steiner's pedagogy and his agricultural methods in an attempt to provide both meaningful agricultural education and proper nutrition. I will discuss these institutions and their methods in the course of this study.

In September of 1924, Rudolf Steiner fell ill. Although he continued to write, his public lecturing career ended on 28 September alter some six thousand addresses. He died in Dornach, Switzerland on March 30, 1925.[46]

**Footnotes for Chapter Two**

1. Russell Davenport, *The Dignity of Man* (New York: Harper and Row, 1955) 335.
2.Stewart C. Easton wrote *The Western Heritage* (New York: Hoyt, Rinehart & Winston, 1961) and two other books on the history of Western civilization. Easton also has been active in the curative education work begun by Steiner, and he has written extensively on anthroposophy.
3. Stewart C. Easton, *Man and World in the Light of Anthroposophy* (New York: 1975) 9. This work is an attempt to integrate the multifaceted work of Steiner and that which has grown from it into a single volume. It provides a very good introduction. For a more detailed look into specific cultural institutions, see the articles in John Davy, *Work Arising from the Life of Rudolf Steiner* (London: 1975) and in A.C. Harwood, *The Faithful Thinker* (London: 1961).
4. Rudolf Steiner, *The Course of My Life* (New York: 1951). In addition to this autobiography, the best biographical sources are: Johannes Hemleben, *Rudolf Steiner: A Documentary Biography* (East Grinstead, Sussex: Henry Goulden, Ltd., 1975); Guenther Wachsmuth, *The Life and Work of Rudolf Steiner* (New York: Whittier Books,1955) the most detailed source for the years from 1900 to 1925; and A.P. Shepard, *A Scientist of the Invisible* (London: Hodder and Stoughton, 1954). A very good short work with good illustrations is Frans Carlgren, *Rudolf Steiner* (Dornach, Switzerland: Philosophic-Anthroposophic Press, 1972). Perhaps the best single biography is Stewart Easton, *Rudolf Steiner: Herald of a New Epoch* (New York: 1982).
5. Steiner, Ibid., 12, 11.
6. Ibid., ii.
7. John Davy, "Rudolf Steiner," in Davy, *Work Arising*, 13.
8. Steiner, *The Course of My Life*, 24.
9. Hemleben, *Rudolf Steiner*, 19-20.
10 Steiner, *The Course of My Life*, 26.
11. Davy, "Rudolf Steiner," 14; Steiner, Ibid., 42-43.
12. Hemleben, Rudolf Steiner, 37-38. The essays by Steiner are gathered into the volume *Goethe the Scientist* (New York: 1950).
13. Hemleben, Ibid. , 39, 44-49. See for example Steiner's *Friedrich Nietzsche* (Blauvelt, New York: Rudolf Steiner Publications, 1960).
14. Hemleben, Ibid., 61. The title, *Die Philosophie der Freiheit*, has been translated as *The Philosophy of Spiritual Activity* (Englewood, New Jersey: Rudolf Steiner Publications, 1963) and as *The Philosophy on Freedom* (London: 1964). The literal translation of freiheit is "freehood'. "Spiritual activity" is the English equivalent given by Steiner himself.
15. Steiner, *Philosophy of Freedom*, xxii.
16. Eddington cited in Lehrs, *Man or Matter*, 70.
17. Steiner, *Philosophy of Freedom*, xxix.
19. Ibid., xxx.
19. Ibid., 14.
20. Ibid., 85.
21. Ibid., 87.
22. Rudolf Steiner, "The Origins of the Theory of Metamorphosis," published as the introduction to Goethe, *The Metamorphosis of Plants* (Springfield, Ill.: Biodynamic Farming and Gardening Assn., 1974); Stewart C. Easton, *Man and World*, 276.
23. Rudolf Steiner, *Knowledge of Higher Worlds and its Attainment*, 3rd ed. (New York: 1947) 1.
24. Easton, *Man and World*, 277; Oliver Matthews, "Religious Renewal," in Davy, ed. *Work Arising*, 217.
25. These books are all in print and are available from the Anthroposophic Press.
26. Hemleben, *Rudolf Steiner*, 158-159.
27. A. C. Harwood, "Threefold Man," in Davy, *Work Arising*, 25.
28. Hemleben, *Rudolf Steiner*, 110; Easton, *Man and World*, 297-300.
29. Rex Raab, "Architecture--Buildings for Lie," in Davy, *Work Arising*, 61-76. See the beautiful volume by Steiner, *Baugedanke des Goetheanum* (Stuttgart: Verlag Freie Geistesleben, 1958) text trans. Also Hagen Biesantz and Arne Klingborg, *The Goetheanum* (New York: 1981).

30. Rudolf Steiner, *Von Seelenrätseln*, trans. and . by Owen Barfield as *The Case for Anthroposophy* (London:1970); Guenther Wachsmuth, *The Life and Work of Rudolf, Steiner*, 326-327. A full treatment can be found in Steiner's most comprehensive work on psychology, *The Study of Man* (London: 1966).
31. Werner Glas, *Speech Education in the Primary Grades of Waldorf Schools* (Wilmington, Del.: Sunbridge College Press, 1974) 15. This book contains a very concise description of Steiner's concept of the threefold human being, and it takes into account the later work by Husemann and Kolisko. For the agricultural aspects of this image, see the excellent work by Martin W. Pfeiffer, *The Agricultural Individuality: A Picture of the Human Being* (Kimberton: Biodynamic Literature, 1990).
32. Ibid., 15
33. Ibid., 16.
34. A.C. Harwood, "Threefold Man," 31.
35. Glas, *Speech Education*, 17; Harwood, "Threefold Man," See especially Steiner, *The Study of Man*, 71-83.
36. Owen Barfield, in his introduction to Steiner's *The Case for Anthroposophy*, 21-22.
37. Harwood, "Threefold Man," 36.
38. Steiner, *The Education of the Child in the Light on Anthroposophy*, 2nd ed. (London: 1975) 24-25; 41-42.
39. Ibid., 42-43; Easton, *Man and World*, 392-398.
40. A good discussion can be found in Glas, *Speech Education*, 41-67. Wachsmuth. *The Life and Work of Rudolf Steiner*, 545, 420-424, 469-570. The lecture course to farmers is published as *Agriculture*, 3rd. ed.(London: Biodynamic Agricultural Assn., 1974).
42. E. Pfeiffer, *Bio-dynamics* (Springfield, IL.: Biodynamic Farming And Gardening Assn., 1948, 1956) 30.
43.H. H. Koepf, *What Is Bio-dynamic Agriculture?* (Springfield: Bio-dynamic Farming and Gardening Assn., 1976) 15-16.
44. Ibid., 14-15.
45. H. H. Koepf,'Three Lectures on Biodynamics," *Biodynamics*, 88(Fall 1968) 20-21.
46. Wachsmuth, *The Life and Work of Rudolf Steiner*, 584.

# PART TWO:

# AGRICULTURE AND THE DEVELOPMENT OF HUMAN CONSCIOUSNESS

## CHAPTER THREE:

## FROM PREHISTORY THROUGH GREECE AND ROME

*Agriculture and animal husbandry have provided the base for all subsequent technological progress, including the current Industrial Revolution, and also the base for the way of life of all of the civilizations that have risen and fallen to date.*[1]

*Arnold Toynbee*

*Since the roots of our (ecological) trouble are so largely religious, the remedy must also be essentially religious, whether we call it that or rot. We must rethink and refeel our nature and destiny.*[2]

*Lynn White, Jr.*

*And here we may conclude that the theocratic and patriarchal element with its roots in the East, can really only produce something consonant with an agrarian system, with a social organization based on a cultivation of the land, on an arable economy.*[3]

*Rudolf Steiner*

### Cultural Epochs: An Introduction

This chapter considers the historical development of agricultural education as it illustrates the evolution of human consciousness. I have used the work of Rudolf Steiner, particularly his scheme of cultural epochs, and I have supplemented
Steiner's original indications with the work of his followers such as Guenther Wachsmuth, D.J. van Bemmelen, and Stewart Easton. I have attempted to show, at least in a rudimentary way, how Steiner's concepts of human evolution in history provide a helpful frame of reference from which to view agricultural education. I used the work of contemporary agricultural historians to supply much of the data, since Steiner never addressed himself directly to this question. He wrote and spoke many times about the development of consciousness and how cultural change results from changes in consciousness, so I have applied these ideas to agriculture and how it was taught.

For Steiner, the physical evidence of civilization's ebb and flow—the cultural historian's domain—are footprints left by the development of human consciousness. The ideas, the thoughts that create cultural change are the "fossils" of human thinking. Rather than viewing the evolution of humanity as a march of progress from ignorant to intelligent, from simple savages to wise modern scientists and lawyers, Steiner cautions us to realize that there is a subtle balance at work in the evolution of consciousness. As one faculty ebbs, another one becomes dominant. As the cosmic wisdom of ancient cultures was forgotten or discredited, the intellect became stronger and gradually became accepted as the sole faculty by which knowledge could be acquired. For Steiner, then, one could see history equally well as the decline of spiritual abilities and the increase of intellectual knowledge, or the decline of religious superstition and the increase of scientific clarity. It is not a question of deciding whether one or the other view is solely correct, but of learning to discriminate between them, to see which faculty is increasing or decreasing on the fulcrum of evolution. Rather than examining only the development of ideas, we need to learn to see how the ideas represent fundamental changes in such faculties of consciousness as perception and cognition.

In the march of humanity's cultural development from the Eastern agrarian theocracies to the modern Western industrial state, we can see the background of our current ecological and agricultural problems. My focus on agricultural education must be seen in this light.

Rudolf Steiner considered all cultural life to be a result of human consciousness—how a person thinks and feels, how he perceives himself and his total environment. He said that humanity has gone through definite stages in the evolution of consciousness, stages which parallel the rise and decline of certain major civilizations. Yet these stages of consciousness and their cultural manifestations are not only to be abstracted from the earthly remains of past civilizations; rather, they are connected equally to the workings of the universe, to the planets and the fixed stars. In a certain sense, one could understand the history of culture by looking out toward the cosmos as well as by looking within, and this reveals one of the hallmarks of Steiner's work—his endeavor to show that man and universe are inextricably wedded into a whole, a unity. For Steiner, changes in the inner life of man always are reflected in the astronomical march of the sun and planets through the twelve zodiacal constellations.

In addition to this astronomical correlation, a corresponding psychosomatic change can be observed in man, and this assertion marks another characteristic of Steiner's thought: the central place he accords humanity in earthly evolution. Steiner's monistic outlook, while he discussed a staggering diversity of phenomena in every facet of life, remained firm in its assertion of cosmic, earthly, and human unity. From this outlook resulted the views delineated in this chapter, the psychological-biological basis for teaching specific cultural epochs to elementary school children I discussed in the previous chapter, and ultimately all Steiner's work.

Perhaps the most striking similarity between human and cosmic rhythm occurs in the relation between man's breathing and the Platonic year.[4] Breathing normally, the human being takes eighteen breaths per minute, 1,080 per hour, or 25,920 breaths per day. The spring equinox, when night and day are of equal length, and the first day of Spring, occurs slightly earlier each year: the sun rises slightly earlier each March 21st. Astronomers generally assert that it would take approximately 25,800 years for the sun to move through the twelve zodiacal constellations and rise again at exactly the same time. Rudolf Steiner said it takes exactly 25,920 years. The Psalms give the average, ordinary lifetime of a human being as three score and ten years, in which there are 25,920 days. Thus the following proportions become apparent: a single breath is to a day (25,920 breaths) as a day is to a lifetime (25,920 days) and as an earthly year is to a cosmic year (25,920 years). Plato spoke in the *Timaeus* of this cosmic year as the "perfect" year, so it is often referred to as the "Platonic" Year, and he derived his mathematical knowledge from the mystery school of Pythagoras, about which we will speak later.

Steiner said that a cultural epoch lasts approximately 2,160 years, which is one-twelfth of a Platonic year, one-twelfth of the time the sun takes to move through the twelve constellations of the zodiac. This precession of the equinoxes causes the first day of Spring to proceed "backwards", or westerly, the opposite direction from its annual easterly path. While the vernal equinox now occurs in Pisces, during Roman times this occurred in Aries; and in Egyptian times the equinox came in Taurus. Many legends and myths point to "this supra-historical reality."[5] The Minoan and Egyptian cultures worshiped the Bull. Jason and the Argonauts searched for the Golden Fleece. Abraham slaughtered a ram, and many called Christ the Lamb of God. The early Christians also used the fish as their symbol, showing their awareness of the approach of the Piscean age and the ending of the age of Aries, represented by the symbol of the Ram.

Like the ancient peoples who were cognizant of such cosmic-human connections, notably the Chaldeans, Rudolf Steiner regarded the beginning of a cultural epoch as the time when the sun passed the mid-point of the constellation. He told of four such cultural epochs of 2,160 years that preceded our own modern era, and he termed these the Indian, Persian, Egyptian, and Greco-Roman epochs.

Steiner decried the modern tendency to reduce all phenomenological data to the simplest explanation. His schema of human evolution, while it can be presented in its outline as I am attempting here, Steiner sculpted as an edifice of great complexity, which he approached from many different points of view. His concept of cultural epochs should be seen as a tool for understanding diverse cultures and their development, particularly for understanding how human faculties such as perception and cognition have evolved. The beginning and ending dates of the epochs, then, are only signposts to alert us to certain changes. For example, some developments of Greek civilization began earlier, or were refinements of Egyptian practices, which in turn go back even further. Thus the cosmic proportions first written down by ancient Chaldeans from their observations of the heavens found their way into the proportions of Egyptian temples, and later into Greek temples, finally coming to rest in the great cathedrals such as Chartres. Yet plainly, the Acropolis buildings represent a qualitative leap from the pyramids or the temples at Karnak, just as Chartres differs significantly from the Erechtheum. The underlying proportions may be strikingly similar, as Steiner and others have shown, but surely the consciousness of the builders, and the artistic and psychological effects they hoped to achieve on those who entered their structures, was quite different.[6]

Astronomically, the reader should be aware that the constellations are of unequal size. As shown above, a cultural epoch begins when the sun passes approximately halfway through the constellation, so that we arrive at epochs of equal length. Yet great cultural leaders who give the chief cultural impetus to the coming age often appear just as the sun moves into the new constellation. For example, Zarathustra first appears in about 6400 B.C., when the sun moved from Cancer into Gemini; he was most instrumental in founding the second cultural epoch, called the Persian epoch, in which agriculture arose. Hermes, or Thoth, the legendary founder of the Egyptian civilization, and whom Steiner asserted was a real person, appeared in Egypt about 4200 B.C. (when the sun moved into Taurus), but the Egyptian epoch really begins in 2907 B. C. Finally, Moses, who gave such a formative impulse to the Judeo-Christian culture, was born in Egypt about 1200 B.C., when the sun entered Aries. The real result of his work can be seen permeating the fourth epoch, which began in 747 B.C., and ended in 1413 A.D., in which the intellectual study of the law and the idea of lineal, as opposed to cyclical, history became dominant.

Another point to remember is that developments characteristic of one epoch often appear as prototypes in the previous age. One example of this is the improvement in agricultural techniques that took place in the Middle Ages. While the argument could be made that this is a characteristic of the so called Agricultural Revolution of the sixteenth and seventeenth centuries, closer examination shows it to be a foreshadowing of later times, and certainly of nowhere near the magnitude of the changes brought by the scientific examination of farming of the eighteenth and early nineteenth centuries, and the mechanization of farming in the late nineteenth and twentieth centuries.

I am aware that the following discussion leaves out some great civilizations, most notably China and the American Incas, both of which developed subtle agricultural techniques. Steiner said very little about these societies, but I believe that my discussion of the centralized, theocratic, agrarian civilization of Mesopotamia and Egypt will cover many of the points raised by a closer examination of China, the Mayans, and the Inca. The other point to be made here is that we are considering a specific spiritual-cultural stream, which Steiner called the Sun-mystery. A quite different history would be written if we followed the development of another mystery stream, such as the Saturn mysteries of the Native Americans or the Jupiter stream.[7]

I ask my readers to remember that the thinking that has created the world economic system and mosern science, with their devastating effect on world agriculture developed first in Europe. I believe that this thinking is the root both of the cause of our problems and the beginning point of the solution. We can turn profitably neither to the cultural-religious fanaticism that looks to the past, nor to cultural chauvinism (such as the "Eurocentric" view. The Sun mystery of individualized consciousness is in no way European or American nor the property of any other culture or nation. It is cosmic and inevitable. And it is just the sort of larger evolutionary picture presented here that can help us peer beyond the present moment to see the longer threads of development.

I have gone into such detail as a warning not to take Steiner's work as a simplistic formulation, but rather to realize that Steiner meant it to provide some order to the ever increasing data being recovered by modern historians, as well as illuminating the reality he saw behind myth and legend. After making a study of Steiner's evolutionary thought, the cultural historian Theodore Rozsak surmised that one could properly appreciate the profundity of it only by following the meditative path which Steiner gave for those who wish to achieve some personal verification of his spiritual research.[8] Steiner himself thought otherwise: that the results of his research could be judged for their truth by anyone willing to read them with an open mind and a healthy understanding. Certainly a meditative approach to Steiner's work, in the opinion of this writer, can only help one achieve the living thinking necessary to comprehend it fully; yet anyone with a strong and good will can gain much from Steiner's ideas and can apply his indications to many endeavors other than the study of history.

## Individualization and Mystery Wisdom

I have selected two main themes from Steiner's idea of the evolution of consciousness. One is the increasing individualization of the human being, and the resulting changes this growing sense of self brought to cultural life. Steiner maintained that this process caused many of the social and psychological problems of Western civilization, but also brought most of the advances of the past centuries. We shall see that for every gain in consciousness, as one or another ability such as abstract thinking is developed, corresponding faculties are lost for a time.

The other theme is Steiner's teaching of certain great leaders of men, whom he called "initiates," who pioneered each step forward in the history of human consciousness, and who gathered groups of people in order to teach them the things necessary to be learned, which later would become faculties or abilities possessed by many people. Examples of such men are Zarathustra, Hermes, Pythagoras, Plato, and Christian Rosenkreutz. Several of these men are somewhat shadowy figures of legend, yet Steiner asserted many times that they did indeed exist, and he gave detailed descriptions of their knowledge and the techniques and results of their teachings.

As the human race became more individualized, the centers of initiation, which Steiner called "mystery centers" or "mystery schools", receded more and more into the background. In the Persian and Egyptian epochs, and certainly before them, all cultural life centered around the temple, out of which the initiates and their pupils taught and ordered the social, economic, and religious life of the people. Increasing individualization paralleled the shift of the cultural pioneers from East to West, and also proceeded hand in hand with the gradual split of a once harmonious, unified cultural life into science, religion, and philosophy, ultimately leading to the fragmentation seen today in modern Western culture. In many lectures Steiner reiterated this story of the growing sense of self and the movement of the location of greatest cultural activity, of the center of the growth of new faculties, from East to West, from paleolithic India to Persia, Egypt and Chaldea, Greece and Rome, and finally to Europe and the United States.

Along with the passing into the shadows of the mystery centers came a development well known to cultural historians—a change from mythological explanations of the world of nature and humanity's role in it, to philosophy, and finally to the modern scientific world-view. This "mythopoeic" consciousness, to borrow Henri Frankfort's phrase, was a picture-forming consciousness; human beings thought in pictures rather than in the concepts we employ today. Frankfort and others have traced the development of pictorial explanations of natural and spiritual phenomena from mythology to the beginnings of philosophical thinking in Greece, and Frankfort's book *Before Philosophy: The Intellectual Adventure of Ancient Man*, is most helpful in this matter.[9]

Rudolf Steiner, by virtue of his clairvoyant powers, claimed that the origins of mythology lay in the direct perception of spiritual beings and forces that are the true causal agents behind the physical world we perceive with our senses. The loss of direct spiritual perceptions that played such a great role in the thinking of the ancients, such as the Hindu doctrine of the Kali Yuga (the Dark Age), of the Egyptian idea of the "Veil of Isis," or the Judeo-Christian concept of the "Fall of Man," Steiner regarded as not merely mythological stories, but as correct pictures of reality. He saw modern conceptual thinking as almost a polar opposite to the dreamy clairvoyance with which ancient peoples apprehended the world. For our thinking to achieve its present power, a corresponding loss of the old clairvoyance occurred, until by the nineteenth century, the clairvoyant perception of the spiritual world existed in very few people in the western world.

Even today, people in the less developed parts of the world tend to retain remnants of the old powers much more than in more "civilized" areas. Familiar examples bf this are country people in Ireland and Wales, and the Scottish Highlands of even two generations ago, not to mention the Indians of Mexico and the Bushmen. I personally know an old man in Fort Bend County, Texas, who can diagnose with great accuracy the most varied animal diseases merely by looking at the animal. Yet he can barely read and write. When asked how he knows this, he says "Just look in their eyes; can't *you* see it?"

**The Indian Epoch, 7227-5067 B.C.**

If we look back to ancient India, to the epoch between 7227 B.C. and 5067 B.C., Steiner speaks of a very advanced civilization, in which nearly all people were clairvoyant, and where great spiritual teachers, who have come down to us through legend as the Seven Holy Rishis, taught the populace. A pale reflection of their teachings exists in the priestly poems called the Vedas, but it must be remembered that these poems, still of almost unearthly beauty, were written down nearly three thousand years after the Indian epoch ended, long after the glories of spiritual insight had begun to fade. The Indian civilization which comes most readily to mind today, which gave to the world the Bhagavad Gita and the Upanishads, arose in the third cultural epoch (2907 B.C.-747 B.C.).

The Indian epoch lies far beyond the existence of external historical records. Steiner said it was a culture totally permeated by supersensible wisdom, in which for most people

> . . . little preparation was needed to arouse in them the scarcely extinct faculties that led to a perception of the supersensible world. For the longing for this world was a fundamental mood of the Indian soul. The Indian felt that in this supersensible world was the primeval home of mankind. From it he was removed into a world that is now revealed only through the perceptions of the outer senses and grasped by the intellect bound to these perceptions. He felt the supersensible world as the true one and the sensory world as a deception of human perception, an illusion (Maya).[10]

Not only are there no external historical records of this time, but Steiner also held that the Rishis taught their pupils supersensibly. There was no physical transmission of language as we know it today; rather, knowledge passed directly from the teacher to the people. When the planets were in a certain conjunction, the people would gather before one of the Rishis. The teaching would stream from his eyes into the eyes of the pupils. We may wish that education could be given so directly today! One of the chief teachings concerned the great Sun Spirit, known as "the Concealed One." To these people, the supersensible world appeared more real than the physical world; the sensory world seemed dim and dreamlike by comparison. Agriculture seems to have played no role in the lives of the ancient Indians. Indeed, there was little need for it because their psychic abilities gave them such a deep insight into the natural world that they were able to procure what foods they needed by gathering herbs, vegetables, and fruits; perhaps a remnant of these abilities can be seen in the profound knowledge of the chemical properties of plants which aboriginal peoples as the Bushmen of the Kalahari still possess. A personal student of Steiner, the physician Friedrich Huseman, wrote:

> These early forms of consciousness were of such a nature that the connections between the human soul and the organism and environment were directly visible to perception, whereas today we are compelled to use observation, experiment, and thought. Only thus is the astounding, insight of "primitive races" in the realm of natural concepts and medicine to be understood.[11]

As the Indian epoch progressed, perceptions of the physical world grew sharper, and clairvoyance dimmed. The dreamlike awareness of the physical world became brighter, while the direct perception of the spirit became much more dreamy. The state of consciousness by which the ancient people apprehended the spiritual world to a state between our waking and sleeping Steiner called "intermediate consciousness states."[12] This fundamental shift of perception, he said, shows itself in the paleolithic cave drawings (a variant of which the incredibly skillful Bushmen and aboriginal Australian artists practiced well into our century). Steiner maintained that these drawings were in fact centers of instruction where young hunters learned to see clearly the physical outlines of animals, In short, the drawings trained faculties of perception.[13]

Our dream consciousness is a last survival of the ancient clairvoyance and this partly explains why in later ages dreams were thought to be a path toward foretelling the future or otherwise penetrating to spiritual realms. Perhaps, too, we can see from the imagery of our own dream-life how it was that the ancients lived so deeply in imagery, in their imagination. Many myths, then, are pictorial representations of direct spiritual realities; it is our inability to perceive these realities, Steiner said, that leads modern mar to see these pictures as mere visionary or hallucinatory experiences. Here we should pause to remember Steiner's statement that by enhancing and strengthening imagination, modern people can once again learn to see behind the sense-perceptible world. He gave many exercises toward this end.

Rudolf Steiner emphasized that the older clairvoyance was not a chaotic flow of images, as our dreams often are, but rather was a reliable means for perceiving the spiritual ground of the natural world. He said that

> Just as present-day man is now convinced through his sense-perceptions and intellectuality that his blood is composed of substances which exist without in the physical universe, so was prehistoric man confident that his soul and spirit emanated from that same hidden spirit world which he could discern by virtue of his clairvoyance.[14]

As spiritual perceptions dimmed and physical perceptions grew, humanity began to turn his efforts toward the earth. The stage was set for the birth of agriculture.

**The Persian Epoch and the Beginning of Agriculture**

The origins of agriculture are hidden in myth and legend. When and how were animals domesticated? How did wheat evolve out of wild grasses? How did man learn to plow? Often called the Neolithic Revolution—when man began to leave his nomadic and pastoral ways and take up the more sedentary life of farming and gardening—this tremendous change occurred in the mists of prehistory. Jacquetta Hawkes held that there are very few precise facts about the first steps in the cultivation of wild cereals. "We have absolutely no tangible evidence for this period of tentative transition," she wrote.[15] Historians tend to agree only that the first evidence of domesticated cereals and some of the earliest evidence of animal domestication were found in an area "stretching from Greece and Crete in the foothills of the West to the foothills of the Hindu Kush south of the Caspian Sea in the East."[16]

Admittedly, this is a very large area, and the chronological gap is correspondingly large. Best estimates place the first evidence of wheat and barley growing in communities that kept sheep and goats in the sixth millennium B.C. The next domestications were lentils, flax, peas, and vetch; by the fourth century B.C., farmers had placed the vine, olive, and fig under cultivation. There is some evidence that between the fifth and sixth millennia B.C., farming communities arose between the flood plains of the Tigris and Euphrates rivers,[17] but the record is sketchy at best before approximately 2500 B.C.[18]

How were the cereals domesticated? After making a thorough search through the literature on the subject, J. Bronowski asserted that

> The turning point to the spread of agriculture in the Old World was almost certainly the occurrence of two forms of wheat with a large, full head of seeds. Before 8000 B.C. wheat was not the luxuriant plant it is today; it was merely one of many wild grasses spread throughout the Middle East. *By some genetic accident*, the wild plant crossed with a natural goat grass and formed a fertile hybrid.[19] (My emphasis)

Genetically, this fortuitous event combined the fourteen chromosomes of wild wheat with the fourteen chromosomes of goat grass to produce Emmer, a plant with twenty-eight chromosomes. Emmer is not only much more plump than grass, but its seeds are attached to the husk so the wind can broadcast them.[20]

Then another very unusual thing happened. The emmer, already under cultivation, again crossed with a natural goat grass of fourteen chromosomes to produce a bread wheat of forty-two chromosomes. Hybrids are usually infertile, but a specific mutation of one chromosome made the new bread wheat fertile. While the emmer hybrid would scatter in the wind if the par broke, the new wheat plant was too heavy to do so. How a symbiotic relationship had to arise between man and wheat - wheat ad to be harvested and replanted. This in turn called for new technology: a sickle with a serrated edge to cut the wheat so the ears would not fall down, and a plow to move the earth, which also required draught animals.[21] Civilization was turning from Abel's flocks to Cain's furrows.

Rudolf Steiner presented a somewhat different picture of agriculture's birth, one that accords more fully with myth and legend— the forerunners of history—and one that seems to fit the centralized, theocratic character of the later historical societies of Sumer, Babylon, and Egypt. The conventional theory, in which groups of shepherds just happen to find a fertile hybrid which produces another fertile hybrid, which leads several people to meld the principle of the wedge for cutting the soil to the lever for turning it thereby inventing the plow—this chain of circumstances appears to run counter to all available historical information about the human personality of the time. For when we examine the artifacts of the third epoch's cultures, we find very little evidence of individualization. Even as late as 700 B.C., the arts showed "a tendency to endless repetition of attitudes, and an inability to distinguish individuals save by the most obvious means."[22] Steiner concluded that the Neolithic Revolution did not occur by chance, but was largely the work of one of humankind's great leaders, Zarathustra.

In the oldest section of the Persian scripture, the *Zend Avesta*, called the "Gathas," Zarathustra is called the man "who was the first Priest, the first Warrior, the first Plougher of the ground, who was the first Prophet and the first Teacher."[23] Rudolf Steiner said that Zarathustra gave his followers techniques for the elevation of consciousness, but that this higher state of consciousness did not result in a withdrawal from the earth, but in an attitude of intensified service to the earth. Agriculture became a religious duty.[24]

The work of Zarathustra in Persia during the seventh millennium B.C. began a fundamental shift in human consciousness. In the Persian cultural epoch, the Persians no longer regarded the sensory world as Maya or illusion. As Steiner put it:

> The peoples of this second period had a different task from the Indian. In their longings and inclinations they did not turn merely toward the supersensible; they were eminently fitted for the physical-sensory world. They grew fond of the earth. They valued what the human being could conquer on the earth and what he could win through its forces.[25]

The Persians had a legend in which the Sun God, Ahura Mazda, or Ormuzd, which means "sun aura", whom they saw as standing behind the sunlight, gave the great impetus to their culture by giving a golden sword to their leader, Yima. Ahura Mazda told Yima, "Thou blessed Yima, child of the sun, expand the earth with thy heels and thy hands, split it apart; like wise men expand

the earth by tilling it."[26]
Yima and his followers, under the tutelage of Zarathustra, used the "Golden Blade" to cultivate the soil, and thus the archetypal plow came into use.

Steiner pointed to numerous things in the Zend Avesta which to many modern readers seem only flights of fancy, but which he asserted were actual indications of correspondences between the inner and outer worlds, the human and cosmic worlds—as he called it, the microcosmic and macrocosmic. For example, the twelve Amchaspands and the twenty-eight Izeds whom the Amchaspands rule, Steiner said correspond to the twenty-eight spinal nerves which are subservient to the twelve cranial nerves. Thus what the ancient clairvoyant perceived as spiritual beings and their physical counterparts, the modern scientist sees only as a fact of human anatomy. Ancient peoples had no capacity for intellectual thought, so these truths were given in such imaginative pictures. It was Zarathustra too, who bred the wild grasses into cereals such as wheat and barley in order to provide the physical nourishment to carry the new consciousness.[27]

From Zarathustra, the Persians learned that the Sun-being, whom they perceived clairvoyantly in the sun's aura, would one day descend to the earth and unite himself with mankind. In preparation for that time, humanity was to view the earth as his true home, and to see it as the battleground between good and evil powers, between light and dark. Agriculture was a means to open the earth and admit the forces of light into the dark earth forces.[28] To the ancient Persians, this view of agricultural work seemed the same as their inner striving: both in the fields and in their own souls they tried to aid the cosmic and spiritual forces of Light to spiritualize the dark forces of the earth and the dark side of their inner nature. We will see this cosmic-earthly perspective repeated several times in the subsequent epochs.

### The Third Epoch: Egypt and Chaldea, 2907-747 B.C.

In the second cultural epoch, the Persians' healthy focus of consciousness toward the earth began to pose a problem, one that seems intrinsic to human nature. If the fundamental cast of mind of the ancient Indian was a strong memory of the supersensible world and an intense longing to return to it, so the Persian sought to enter fully into earthly life. Yet each view tended to be one-sided in its own way. The search for balance continued into the third epoch. In a certain sense, the Mesopotamian civilizations, notably Babylon and Chaldea, carried on the Persian approach, while the more inward-looking Indian soul metamorphosed into the Egyptian civilization. The need for balance played a large role in both Mystery centers, and a further cosmic event hampered their efforts: the beginning of the Kali Yuga, or Dark Age.

Ages of spiritual enlightenment and of spiritual darkening occur in five thousand year cycles, a cosmic rhythm known to very many ancient peoples. Steiner used the Sanskrit term, Kali Yuga, for it. He dated the years of the Kali Yuga as 3101 B.C. - 1899 A.D. Steiner characterized the Kali Yuga as a period in which the clairvoyant powers of mankind grew increasingly dim and confused, until by the nineteenth century, only a mere handful of initiates retained the direct perception of the spiritual world.[29] The great majority of people in Western "civilized" nations became increasingly materialistic in their thinking and thus in their world-conceptions, until materialism reigned supreme by the end of the nineteenth century. I cannot think of any culture which has not felt the effects of this age of darkness, with its impetus toward materialism in every aspect of life.

Although this third cultural epoch marks the beginning of recorded history, the fact that most of humanity left the prehistorical era doesn't mean that human consciousness had reached a state comparable to modern mankind. Even the Greeks of Socrates's time were far different than the Egyptians and Chaldeans. Yet the glories of Greece and the imperial splendor of Rome owe much to their eastern antecedents in the evolution of consciousness. Certain characteristics of civilization, from architecture and farming to the most secret rites of the Mystery centers in the temples continue to influence even modern life. Sir Leonard Woolley wrote that "the arch, the vault, the apse, and the dome used in Europe first in Rome, were all used long before by the Sumerians."[30] The Masonic lodges and other esoterically oriented groups continue to use Egyptian rites in their initiations. The Washington Monument is shaped like an Egyptian obelisk, Egyptians widely used cosmetics, and of course embalming is very widespread in Western cultures. The third cultural epoch has left its mark on many aspects of our own times.

The civilizations that arose in Egypt and Mesopotamia were very centralized societies which, for all their great size, still were administered in every detail by religious leaders in the temple. These initiates did not rule merely by tradition and superstition, but:

> . . . we find, at the beginning of historical life, a universal, penetrating wisdom, according to which man directed his life. It was not an acquired wisdom, but it flowed to mankind through revelation, through a kind of inspiration.[31]

The evolution of civilization, Steiner repeated many times, flowed "not from primordial ignorance, but from an era of primal wisdom"[32] The kings and pharaohs all received training in the Mystery centers and, at least at the beginning of the third cultural epoch, were virtually indistinguishable from the priests.

Rudolf Steiner portrayed those civilizations as

> . . . a complete unity. At that time there existed Mystery centers which were also centers for education and culture, centers dedicated at one and the same time to the cultivation of religion, art, and science. For then what was imparted as knowledge consisted of pictures, representations and mental images of the spiritual world. These were received in such an intuitive and comprehensive way that they were transformed into external sense-perceptible symbols and thereby became the basis of cultic ceremonial. Science was embodied in such cults, as was art also; for what was taken from the sphere of knowledge and given external form must perforce be beautiful. Thus in those times a divine truth, a moral goodness and a sense-perceptible beauty existed in the Mystery centers, as a unity comprised of religion, art, and science.[33]

Numerous historians point to this unity of society, and of its influence on the life of man. For example, Jacquetta Hawkes wrote: "The dramatic advance of civilization in the late fourth millennium was led in the name of the gods and through religiously inspired organization."[34] The great Egyptian religious leader, Hermes Trismegistus received his training in the Persian Zarathustrian mysteries, which he re-formulated to fit the Egyptian consciousness. The Egyptian could perceive the spiritual world behind the physical only to a limited extent, but he recognized that spiritual laws worked in the physical world. Hermes taught that if a person works zealously with earthly forces during life according to the aims of the spirit, he would be united with Osiris—the Sun Being—after death.[35]

Yet death loomed increasingly large in the Egyptian mind as the epoch progressed. Where the ancient Indian and Persian wondered "how do I live in an earthly body?," the Egyptian and Babylonian asked, "How do I pass through earthly death?" Embalming practices arose partly in an attempt to preserve man's physical forces from death, a clear indication that these people were losing their knowledge of the spiritual origin of humanity. It illustrates the problem that arose as the perception of the earthly world sharpened and the perception of the spiritual world dimmed.[36]

In the Chaldean-Babylonian civilization, men plunged even deeper into physical existence than in Egypt. Their great star wisdom, a small measure of which remains even today in astrology, became slowly more subject to a materialistic view. Steiner wrote that

> instead of the spirit of the star, the star itself, instead of other spiritual beings, their earthly counterparts were pushed into the foreground. Only the leaders acquired really deep knowledge of the laws of the supersensible world and their interaction.[37]

This appears also in the historical record. P. Campbell Thompson found that by the seventh century B.C., the end of Steiner's third epoch, celestial observation had become much more acute, while attempts to use it to explain the future proceeded by "formal and perfunctory quotations from the old astrological books."[38] Knowledge of the physical world was increasing while the ability to find meaning in it fell first into tradition and then into philosophical abstraction.

The Egypto-Chaldean epoch brought large improvements in agriculture. Needless to say, agriculture was regulated in every way by the temple administrators, who gave the farmers detailed indications from their mystery knowledge. We must realize that an immense distance separated the consciousness of the high initiates from that of the temple workers, who received most of their information in the form of myths, some of which gave specific plowing and planting instructions and directions for proper animal husbandry. It is relevant that the first known artistic representations of the harness are pictured in religious processions, and that a number of Mesopotamian frescoes show priests plowing and performing other arts of husbandry. The fact that the Egyptian farmer could produce three times more food than was necessary to support the population shows that the temple culture took agriculture very seriously.[39]

Thus the deep knowledge of the initiates filtered down into the general populace.[40] For the historian, it is very difficult in the early centuries of the epoch even to distinguish the secular king from "the God:" in examining the records, wrote C. J. Gadd, the historian is "uncertain whether one is on the divine or human plane."[41] The real reason for this, according to Steiner, is that the priest-kings spoke to the people out of their direct revelations of the spiritual world. They were "channels" for the god of the people; in that sense it was impossible, when the king was "spoken through" by the god, to determine whether the human being or the spiritual being was speaking,

The temples operated farms, gardens, orchards, pasture lands, large kitchens and bakeries, breweries, shops, spinning and weaving, and mills. The festivals, quite naturally, came under their aegis. To the workers, who were sometimes slaves, the temple furnished land, plows, cattle, seed, and cultivating implements. The temple supported these workers also during the off-season. Luigi Pareti, in *The Ancient World*, remarked that "every act in the farmer's life was accompanied by the performance of a domestic rite."[42]

As the early Chaldeans charted the heavens, their descendents, took these cosmic-earthly-human proportions and used themm to lay out not only their magnificent buildings, but also fields. Their engineering skills found employment in the elaborate canal systems of Mesopotamia. The irrigation works had to be maintained in spite of a very high siltage rate thus workers constantly had to dig out the canals. The astonishing continuity of leadership provided by a unified culture stemming from the temples can be seen in the over two thousand years of successful irrigation that the Babylonians and Chaldean-Sumerians accomplished. The changing political fortunes altered the lives of the people surprisingly little. Christopher Dawson said that in all essentials, Babylon, in the time of Hammurabi (ca. 1800 B.C.), had reached a level of material civilization which has never since been surpassed in Asia.[43]

In Egypt, the agricultural base provided high production even in the face of a century of political occupation by the Hyksos nomads in about 1700 B.C. The temple culture, it seemed, could withstand nearly any onslaught. Egyptian civilization remained a potent factor in world history for three thousand years.[44]

It appears, based on Rudolf Steiner's research, that Egypt and Babylon fell to the inexorable march of time. Momentous developments in human consciousness followed the focus of evolution west to the Minoan bull-culture, and finally to Greece. Even in Egyptian art, however, we can trace this change, for at the end of Egyptian dominance an indication of individual human features appears in their sculptures and bas-:reliefs. While the full flowering of the human personality took place in Greece and Rome, one mighty figure arises in Egypt and travels toward Israel: the great law-giver Moses.

The significance of Moses for this study is that he was a fore-runner of what Steiner called "ego consciousness." Steiner did not give use the word 'ego' in the same way as modern psychology; rather, he meant "that which says 'I' to itself."[45] Previously, human beings didn't regard their powers of soul ( thinking, feeling, and willing) as internal, but rather as forces working into them from the cosmos: thus their soul forces connected them to the gods—to spiritual beings and forces. With Moses, who appeared about 1400 B.C., we see a harbinger of the new "I"- consciousness. To Moses, the spiritual world did not speak from many directions, but through the center of his being, saying "*Eje ascher eje*" or "I am the I am". Steiner held that this is an archetypal moment in time, in which the spiritual world began to speak to humanity through the *individuality*, the self; and with self-consciousness began the first evidence of the intellectual understanding of nature. Henceforth, Steiner said, God could be found in the spiritual world *and* in the physical world. Moses, like Zarathustra and Hermes before him, came to prepare mankind's descent into the material world, and the "I" or ego into his own self.[46] It would be the inner foundation for the individual, then, to begin his own re-ascent to the spirit. To discuss this further, we must delve into the worlds of Greece and Rome in the fourth cultural epoch.

## The Greco-Roman Epoch To the Time of Christ (747B.C. - 100 A.D.)

*But in the last analysis all external history is dependent on thinking, and what he achieves in history man produces from his thoughts, together with his feelings and impulses of will; and therefore, if we want to find the deepest historical impulses, we must turn to human thinking.*[47]

*Rudolf Steiner*

*Most of the poor peoples of the earth are poor mainly because their ancestors wasted the natural resources on which present generations must live.*

*The fundamental cause for the decline of civilizations in most areas was deterioration of the natural-resource base on which civilization rests.*[48]

*Vernon Gill Carter and Tom Dale*

In contrast to the preceding epoch, in which humanity lived largely in its feeling life, and in which its consciousness functioned mostly in picture-forming—in mental pictures—rather than in conceptual thoughts as we know them, the fourth epoch brought the beginnings of intellectuality, of intellectual thinking. Steiner gave this new soul-quality the name "intellectual soul". The merest familiarity with Greek sculpture and architecture reveals a profound change from Egypt and Mesopotamia. Philosophers such as Plato and Aristotle, not to mention later figures like Albertus Magnus and Aquinas, can scarcely be imagined in Egypt or Babylon. Both the human personality and the intellect come of age in this epoch, a fact which hints of their intimate relation, for one cannot very well form abstract thoughts about the world without both a fairly strong sense of self and the corresponding ability to distinguish between oneself and the larger world. In a sense the birth of philosophy and of the individual personality are the positive results of the "fall of man" from his supersensible origins into a more earth-bound consciousness. The waters of the River Lethe, the mythical River of Forgetfulness, caused the Greek to forget his spiritual home of the life before birth and after death; perhaps we can take this also as a picture of the metamorphosis of the ancient dream-like clairvoyance into fully conscious thinking.

In that sense, then, the fourth epoch witnessed a gradual lessening of man's feeling that the forces and beings of the spiritual world worked through his own being, in his own body, as his thoughts and feelings. To Moses, it was God who spoke to him in the core of his being—through his "I"; to the Greek, it was the Gods who "thought" his thoughts in him. By medieval times, Aquinas and others felt that man's thoughts were largely his own, and he could only hope that these thoughts did no run counter to church dogma. Thus the situation at the beginning of the epoch had become nearly reversed by the end of it. Direct perception of spiritual reality gave way to tradition and dogma. As Rudolf Steiner said:

> Thus we must conceive the source of the primeval wisdom as a spiritual life of rich abundance which becomes impoverished as evolution proceeds, and when at last it reaches the Western world it provides the content of religious creeds. Therefore men who are by then fitted by nature for a different epoch can find in this weakened form of spiritual life only something to be viewed with skepticism.[49]

## Greece and Rome

The philosophical and artistic heights of Greece and the legalistic and engineering skills of Rome are sufficiently known, so we quickly may pass on to elements of the epoch more germane to this study. Steiner emphasized the position of Greece, and Rome to a lesser extent, as a crossroads between East and West, where people from all over the ancient world had gathered; where

> there were men who possessed, as a natural faculty, the heritage of ancient clairvoyance, and there were some who were able to attain to it with comparatively little training.[50]

The ancient spiritual wisdom permeated totally the Greek culture, less so the Roman world. The human personality flowered in Greece, particularly in Athens, where Socrates's life formed a powerful archetype of a future human being who could make moral decisions and seek philosophical truths out of his own being. While the early history of philosophy is widely known and has received admirable discussion from Henri Frankfort and many other writers, the extraordinary degree to which these philosophers owed their ideas to the spiritual teachings of the Mystery centers, such as Eleusis and Ephesus, is sometimes overlooked. Rudolf Steiner devoted much effort toward illuminating this relationship and, due to the direct relation he saw between these teaching centers and the evolution of consciousness, I will attempt to cover the Mysteries in some detail in the next chapter.

I believe the emergence of the individual in the fourth epoch, while it progressed only gradually over the centuries, exhibited many of the strengths and weaknesses intrinsic to its development during the years under examination here. Just those qualities that allowed Socrates to stand firm in the face of the Athenians, that brought forth the first democracy in Athens and the Roman concept of citizenship —these qualities opened themselves to the most blatant abuses, perhaps the greatest being the megalomania of the Caesars such as Caligula and Nero.

Several developments of significance for agriculture arose in Greece and Rome, attributable at least in part to the growing sense of self and the increasing power of the intellect. Discussing the serious soil erosion of Attica, Edward Hyams said

> Moreover, it is important to realize that this destruction was also psychologically possible: the natural environment of primitive, native societies

is, to some extent, protected against its human members by the mythology and traditions which these men have accumulated in their contacts with trees and animals and herbs, and which each generation inherits as a lore and a feeling. But the sophisticated intellectuals who are the men of high civilizations, products of their particular "Socratic revolution; uninfluenced by Orphic feeling, are not inhibited in their assault upon a soil community. They do not feel themselves to be merely members of it, even though dominant ones, but are outside nature, God's tenants, given a free hand with the landlord's property.[51]

The Greeks, and also the Romans, were not the first people to damage their soils—the Indus valley civilization of Harappa and Mohenjo-Daro had deforested large areas of Sind much earlier—but the emerging intellect allowed them to do it more thoroughly. Where it took several thousand years to denude the lands of the Fertile Crescent, the Greeks managed their environmental destruction in several generations.[52]

Agriculture in Rome and Attica led to a progressive soil exhaustion. The soil of Latinum possessed more natural fertility than the Greek soils did, and the Roman engineers devised a number of very clever aqueducts and dams to preserve the falling water supply, due to excessive deforestation.[53] The decreasing ability of native soils to support increasing population led to two different solutions. The Athenians developed a maritime hegemony and colonized several areas of the Mediterranean. Rome began the conquests which led to empire. As these tactics are quite similar to modern empire-building, such as the nineteenth century British Empire, perhaps it is not surprising to find a modern ring in ancient descriptions of the agricultural waste that led to the empires. Two examples will follow. In the *Critias*, Plato had Critias say:

> what now remains of the once rich land is like the skeleton of a sick man, all the fat and soft earth having wasted away, only the bare framework is left. Formerly, many of the present mountains were arable hills, the present marshes were plains full of rich soil; hills were once covered with forests, and produced boundless pasturage that now produce only food for bees. Moreover, the land was enriched by yearly rains which were not lost, as now, by flowing from the bare land into the sea; the soil was deep, it received the water, storing it up in the retentive loamy soil; the water that soaked into the hills provided abundant springs and flowing streams in all districts. Some of the now abandoned shrines, at spots where former fountains existed testify that our description of the land is true.[54]

The Romans exhausted the land of Latinum by the second century B.C., and then turned the lands of Sicily, North Africa, and other provinces into Roman granaries. In doing so, they severely damaged the soils of these areas. By 250 A.D., the damage had proceeded far enough for St. Cyprianus, Bishop of Carthage, to say:

> You must know that the world has grown old and does not retain its former vigor. It bears witness to its own decline. The rainfall and the sun's

warmth are both diminishing; the metals are nearly exhausted; the husbandman is failing in his field . . . springs which once gushed forth liberally, now give barely a trickle of water.[55]

The intellect made its mark on agriculture, and several notable writers took up the subject, perhaps the greatest of whom was Mago the Carthaginian, whom Edward Hyams considers the first person "to apply an essentially urban, non-attached analysis to soil problems, although the work is not 'scientific' in the most pernicious sense."[56] It could not, according to Steiner's ideas, have been truly scientific in the sense of a von Liebig of the nineteenth century, since men of classical times could not stand fully detached from nature; thus Mago was not able to see clearly *both* his observations of nature *and* his thoughts about the observations. He could not divorce sufficiently his thinking from his perceptions. Nevertheless, while they could think and observe very well indeed, many people have commented on the excessive generalization of most ancient natural philosophers. But their works had a very marked influence even through the Middle Ages, where most medieval agricultural works showed a deep debt to these men, not least to the "organizing, one might even say rationalizing, spirit of the Roman agricultural writers."[57] Among the leading classical authorities were: Theophrastus, Aristotle's student, who wrote a nine volume work with a systematic description of agricultural plants; Cato's *de Agricultura*, Varro's three volume *de Re Rustica*; Columella and Virgil.[58]

The Roman and Greek farmers made significant technical progress, and for all the damage they caused by intensive commercial farming, they laid the basis for the medieval improvements in technology. They developed an impressive diversity of methods for soil improvement, including the use of legumes, green and animal manures, proper plowing, dry farming and mulching. The Romans began development of the wheeled plow, which would revolutionize agriculture in the Middle Ages, and they even invented a primitive reaping machine. A lack of sufficient manures and fodder, and failure to move from fallowing to rotation hampered further progress, yet some changes from the two-field and fallow system to a three-field rotation did occur, both in Rome and Greece. Lucerne was imported from Persia to Greece and thence to Rome for a fodder crop to cover the fallow field.[59]

By the time of the birth of Christ, however, signs of decadence had appeared. Athenian democracy gave way to dictatorship, and finally their maritime empire lost influence to Rome and Carthage. The Punic Wars, particularly the second, in which Hannibal laid waste to Latinum for sixteen years, began social changes which soon altered the face of the Republic forever. Soon the smaller landowners lost out to the large latifundia, or plantations, manned by slave labor, mainly Carthaginian. With most of the small farmers, the backbone of the Republic, seen in such a figure as Cincinnatus, pressed into military service, a pattern of exploitative farming ensued, and the peasantry, who had the old intuitive relation to the land and its crops, found their lives quite disrupted.

Roman society became more and more fragmented. No longer did a clear wisdom flow from the oracles and temples into the culture. According to Rudolf Steiner, even the old Mysteries had become largely corrupt and decadent. In the oriental theocracies, men attained their sense of self through the Mystery centers and after intense spiritual preparation, such as in the rigorous yoga training of the Hindus. Now human beings were born into a selfhood with little preparation, with much less spiritual guidance from their culture. The Sophists would never have arisen in a thoroughly spiritualized culture; without a strong sense of self, a Socrates could never have challenged a decaying culture's thinking.

Rudolf Steiner said that as the oriental unity of society gave way to the legalistic, commercial society of Greece and especially Rome, conflicts among people had to arise. The oriental spiritual goal was to attain a sense of self through the Mysteries, culminating in the phrase from the Eleusinian mysteries, "Man, know thyself". The Roman, for example, having already attained this sense, began to wonder about totally new concerns, perhaps the chief of which was the thorny question of human labor. With little sense of self, the old theocratic cultures had little idea of human labor; rather it was accepted as a part of nature—as a "given" factor. The older cultures thought nothing of the fact that slavery supported most of society. With Rome, the increasingly strong concept of the self forced them to attempt to integrate labor into the social system—a great social question to this day, and one of crucial importance in agriculture.[60]

Steiner wrote and said much about what he considered to be the root cause of the classical world's problems: the decadence of the Mysteries. The following chapter will examine the Mystery wisdom and experiences and their relation to Christianity.

**Notes for Chapter III**

1. Arnold Toynbee, *Mankind and Mother Earth* (New York: Oxford Press, 1976) 590.
2. Lynn White, Jr., "The Roots of Our Ecological Crisis," *Science*, 155:1207.
3. Rudolf Steiner, *The Tension Between East and West* (London: Hodder and Stoughton, 1963) 127.
4. This discussion is taken from Rudolf Steiner, *Man: Hieroglyph of the Universe* (London, 1972); Stewart C. Easton, *Man and World in the Light of Anthroposophy* (New York, 1975) 31-32, 287-289, (hereinafter cited as *Man and World*); and Werner Glas, *The Waldorf School Approach to History* (New York, 1984) 14-16.
5. Werner Glas, Ibid. A very thorough discussion of this will be found in Guenther Wachsmuth, *The Evolution of Mankind* (Dornach, Switz.: Philosophic-Anthroposophic Press, 1961) a pioneering attempt to integrate Steiner's work with the latest information from archaeology, geology, and ethnology. Especially pp. 24-60. Much more work lies ahead for future historians of spiritual history. Steiner limited his research largely to the Sun mystery and its evolution.
6. See, for example, Peter Tompkins, *Secrets of the Great Pyramids* (New York: Harper & Row, 1971); Louis Charpentier, *The Mysteries of Chartres Cathedral* (New York: Avon Books, 1966); and Peter Tompkins, *Mysteries of the Mexican Pyramids* (New York: Harper & Row, 1976), and René Querido, *The Golden Age of Chartres* (New York:1990).
7. D.J. van Bemmelen, *Zarathustra*, I (Zeist, Netherlands: Uitgeverij Vrij Geestesleven, 1968) 1-6; Easton, *Man and World*, 32-33.
8. Theodore Roszak, *The Unfinished Animal* (New York: Harper & Row, 1975).
9. Henri Frankfort, *Before Philosophy* (New York: Pelican, 1951).
10. Rudolf Steiner, *An Outline of Occult Science* (New York: 1972) 230-234. The quotation is found on p. 230. This book is basic to an understanding of Steiner's historical thought, particularly Chapters 4 and 6.
11. Friedrich Husemann, *Goethe and the Art of Healing* (London: 1938). See for example, Laurens van der Post *The Lost World of the Kalahari* (New York: Pyramid, 1971) and *The Heart of the Hunter* (New York: Morrow, 1971).

12. Rudolf Steiner, *Turning Points in Spiritual History* (London: 1934) 50.
13. Ibid. 51
14. Ibid. 133-134.
15. Jacquetta Hawkes and Sir Leonard Woolley, *Prehistory and the Beginnings of Civilization* (London: George Allen & Unwin, 1967) 271-272; see also I.E.S. Edwards "The Early Dynastic Period in Egypt", *Cambridge Ancient History*, I (London: Cambridge University Press, 1971) 51.
16. D. B. Grigg, *The Agricultural Systems of the World: An Evolutionary Approach* (London: Cambridge University Press, 1974) 11; G. E. Fussell, *Farming Techniques from Prehistoric to Modern Times* (Oxford: Pergamon Press, 1965) 1-2.
17. D. B. Grigg, Ibid; see also Edward Hyams, *Soil and Civilization* (New York: Harper Colophon Books, 1976) 37-42.
18. "Prefaces", *Cambridge Ancient History*, 1-2, xxi.
19. J. Bronowski, *The Ascent of Man* (Boston: Little, Brown, 1975) 65; G. E. Fussell, *Farming Techniques*, 2-3.
20. J. Bronowski, Ibid.
21. Ibid., 65, 68, 74; G. E. Fussell, *Farming Techniques*, 3.
22. Sidney Smith, "Senacherib and Esarhaddon" *Cambridge Ancient History*, II, 77.
23. D. J. van Bemmelen *Zarathustra*: II (Zeist: Uitgeverij Vrij Geestesleven, 1968) 24.
24. D. J. van Bemmelen, *Zarathustra:* I, 23; Guenther Wachsmuth, *The Evolution of Mankind*, 102.
25. Rudolf Steiner, *Occult Science*, 234; "A Chapter of Occult History", *Anthroposophical Quarterly* (Spring, 1968) 3.
26. D. J. van Bemmelen, *Zarathustra:* I, 17.
27. Rudolf Steiner, *Turning Points in Spiritual History*, 74-75.
28. Rudolf Steiner, *Occult Science*, 237-238.
29. See Stewart C. Easton, *Man and World*, 34-35.
30. Quoted in Rudolf Steiner, *Turning Points in Spiritual History*, 125.
31. Rudolf Steiner, *A Modern Art of Education*, 3rd. ed. rev. (London, 1972)30.
32. Rudolf Steiner, *Turning Points in Spiritual History*, 122.
33. Rudolf Steiner, *Human Values in Education* (London, 1971) 144-145.
34. Jacquetta Hawkes, *The First Great Civilizations* (New York: Knopf, 1973) 37.
35. Rudolf Steiner, *Occult Science*, 240-241.
36. Rudolf Steiner, *A Modern Art of Education*, 217-218.
37. Rudolf Steiner, *Occult Science*, 241; see also Chapter 4 of Steiner's *Occult History* (London, 1957).
38. R. Campbell Thompson, "The Influence of Babylon", *Cambridge Ancient History*, III, 238-239.
39. Hawkes and Woolley, *Prehistory and the Beginnings of Civilization*, 518; Erich Isaac, "On the Domestication of Cattle", in Shepard and McKinley eds. *The Subversive Science: Toward an Ecology* of Man(Boston: Houghton Mifflin, 1969) 198-197; Reay Tannahill, *Food in History* (New York: Stein and Day, 1973) 70.
40. Rudolf Steiner, *The Tension Between East and West*, 120-121.
41. C. J. Gadd, "The Cities of Babylon," *Cambridge Ancient History*, I,2, p. 128.
42. Ibid. 126-127.
43. Christopher Dawson, quoted in Vernon Gill Carter and Tom Dale, *Topsoil and Civilization*, rev. ed. (Norman: University of Oklahoma Press, 1974) 44. The discussion comes from 41-45.
44. Ibid., 29-32.
45. Quoted in Stewart C. Easton, *Man and World*, 35.
46. Rudolf Steiner, "A Chapter in Occult History" *Anthroposophical Quarterly* (Spring, 1968) 7; *Turning Points in Spiritual History*, 191-207.
47. Rudolf Steiner, *The Tension Between East and West*, 56.
48. Vernon Gill Carter and Tom Dale, *Topsoil and Civilization*, 7, 20.
49. For Steiner's thorough treatment of this point see *The Riddles of Philosophy* (New York, 1972) and *The Redemption of Thinking: The Philosophy of Thomas Aquinas* (London: Hodder and Stoughton, 1956). The quotation comes from Steiner's lecture "Natural Science and Its Boundaries," *Golden Blade* (1962) 8.
50. Rudolf Steiner, *Occult Science*, 241.
51. Edward Hyams, *Soil and Civilization*, 94-95.

52. Ibid., 64-72 for the Indus Valley story; Vernon Gill Carter and Tom Dale, *Topsoil and Civilization*, 134.
53. This discussion was taken from Hyams and from Carter and Dale.
54. Plato's *Critias*, quoted in Carter and Dale, *Topsoil and Civilization*, 105.
55. Carter and Dale, *Topsoil and Civilization*, 131-132. The quotation is from page 114.
56. Edward Hyams, *Soil and Civilization*, 128.
57. G. E. Fussell, *The Classical Tradition in Western European Farming* (Rutherford: Farleigh Dickinson University Press, 1972) 66; the quote is by Charles Parain, "The Evolution of Agricultural Techniques in the Middle Ages", *Cambridge Economic History of Europe* I (London: Cambridge University Press, 1966) 124.
58. K. D. White, *Roman Agriculture* (Ithaca: Cornell University Press, 1970) 478, 246; Luigi Pareti, Paolo Brezzi, and Luciano Petech, *The Ancient World* II-2 (London: George Allen & Unwin, 1965) 420-421.
59. K. D. White, *Roman Agriculture*, 144; G. E. Fussell, *Farming Techniques from Prehistoric to Modern Times*, 31; Pareti, Prezzi, and Petech, *The Ancient World*, II-2, 384-31.
60. Rudolf Steiner, *The Tension Between East and West*, 133ff, discusses this at length.

# CHAPTER FOUR

## MYSTERY CENTERS, MYTHS, AND THINKING

*Happy is he who has seen those Mysteries ere he passes beneath the earth. He knows the truth about life's ending, and he knows that its first seeds were God's giving.*

*Pindar*

*Thrice happy they, who, having seen these rites,*
*Then pass to Hades: there to these alone Is granted life, all others evil find.*

*Sophocles*

*Those occult Mysteries . . . when interpreted and explained have more to do with natural science than with theology.*

*Cicero*[1]

After basing a study of seven thousand years of human history largely on the premise that the key to the development of human consciousness lies in the experiences and knowledge obtained from Mystery centers and their initiates, it is now time to determine, from Steiner's work, what those experiences were and how they were gained. I approach the subject with decidedly mixed feelings, however, because although Rudolf Steiner devoted one of his major books and many lectures to the various Mysteries and their evolution, he tirelessly emphasized that our modern, abstract thinking offers a very poor soil in which to form fruitful ideas about these places and processes. He wrote that:

> Ancient Mystery wisdom is like a hothouse plant which must be cherished and cared for in seclusion. To bring it into the atmosphere of everyday conceptions is to put it into an element in which it cannot flourish.[2]

Or, as Plato said in the *Timaeus*: "Now to discover the Maker and Father of this Universe were a task indeed; and having discovered Him, to declare Him unto all men were a thing impossible."[3] Yet the past two thousand years has witnessed the increasingly widespread, public knowledge of this ancient wisdom, albeit with a decreasing understanding. Steiner felt that the attempt must be made to present a picture of the Mysteries in order to clarify certain aspects of the evolution of consciousness.

This chapter deals with the two streams of Mystery wisdom which met in Greece; the relation of the Mystery centers and their initiates to mythology, particularly to agricultural myths; and Steiner's conception of the role of Christ and Christianity in the Mysteries of Western culture. I have tried to present Steiner's teachings as clearly as possible, but the reader should realize that the treatment here is at best a bare outline of Rudolf Steiner's research, and can in no way substitute for a thorough perusal of Steiner's voluminous works on this subject.

## Mystery Wisdom

Three things emerge from even the slightest study of the ancient mysteries: they were shrouded in the deepest secrecy; they transcended any intellectual means of learning and raised the initiate to a very sublime level of experience; and they concerned the great enigma of death and rebirth. The great poet Aeschylus, when accused of betraying secrets of the Eleusinian mystery rites in his dramas, had to seek sanctuary in the temple of Dionysus, and from there he produced legal evidence that he was not an initiate. Thus he avoided the penalty for revealing such secrets: death.[4] Aristotle said that

> the uninitiated are not directed to learn by intellect, but they are led to have inner experiences which make it possible to bring them into an extra-normal soul condition, provided they have the capacity to experience this soul condition in the first place.[5]

Rudolf Steiner portrayed the central Mystery experience as a "rebirth," in which the normal consciousness of the physical world fell away, and a new, higher consciousness ensued, during which the initiate became cognizant of the spiritual world. The ancient initiations are often thought to have caused the initiates to experience a symbolic death and rebirth. Steiner differs from these authorities in asserting that the initiation experience actually involved a near-death experience. The subsequent of rebirth, then, was more "real" than merely symbolic.[6]

Since the beginning of the fourth epoch in 747 B.C., most people experience the sensory world as "real", and what arises in the soul as "not as real", but rather as a "picture" of reality. After initiation through a Mystery rite, the initiate beheld the world in a totally new way. Intense preparation to bring his sensuality under control, to "purify" his feeling life, had been very much like a preparation for death, to "die" to the physical world and its sensual delights. After the Mystery experience, the world of the spirit, of cognition, of the images of the soul—these now seemed truly real; the sense world then formed a lower reality. Thus the new initiate felt completely reborn into a new life, with new thoughts and perceptions. While one must *experience* this rebirth, which cannot be proven with external evidence, Steiner asserted that non-initiates can understand it if properly communicated.[7]

The reason for the absolute secrecy that surrounded the Mysteries lay in the very real danger that as sensory perceptions lose their "absolute authority" in one's life, spiritual reality may not arise to take their place. The initiate then would have penetrated to that area where all life seems death: the nether world, or Hades. In that realm, either one's identity vanishes, or a new self is met. As Steiner said, "out of spiritual fire, the whole world has been reborn for him." Yet if the initiate did not rise to a new existence, all happiness in life was gone. For he had seen behind the veil of the senses and also behind the myths of his culture; he either arose to a new level of consciousness or remained adrift among the shades of Hades.[8]

Since the Mysteries have an intimate connection with human evolution, naturally they have undergone a parallel developmental sequence. Yet one common thread runs through the Sun Mystery during all the ages, Steiner wrote: the knowledge that behind the myths and behind sensory perceptions stood "the Hidden God." This great being we saw as the "Concealed One" who existed behind the physical sun for the ancient Indian people, and as Ahura Mazda—the Sun Aura—of the Zarathustrian cult. The initiates did not consider this god anthropomorphically, unlike the gods and myths of the general populace.[9]

In ancient times, after humanity had begun to lose his natural clairvoyance, the priesthood

> established contact with the spiritual world through stimulating their unconscious instincts by associating their metabolism with essences of one kind or another. They knew what each plant in nature could develop from their instinctive life by a kind of dream-like spiritualization; they knew that, if this or that plant were eaten, the effect upon their organism was such that they could transport themselves to a particular area of spiritual activity.[10]

These priests thus conducted the aspirant, after rigorous preparation, into a condition, which lasted for three days, in which he lay in a death-like sleep; but his soul looked upon the spiritual world in full consciousness. After this sleep, the initiate remembered his experiences and felt inspired to teach mankind.

Steiner's research led him to say

> . . . that the Mysteries are schools for the training of those faculties which enable the soul of man to have actual vision of the spiritual worlds. They are schools where in a methodical and systematic way, the soul of a man who is sufficiently mature is so guided and trained that he can finally perceive the spiritual world with [spiritual] eyes and ears.
>
> Through all the ages there have been centers for developing the faculty of fully conscious clairvoyance . . . .

The initiate thus learned that the spiritual world is the true origin of humanity, that death is but one of the processes of life, and that true cognition does not depend on bodily processes.[11]

Rudolf Steiner found that the great wisdom of the ancients originated with such initiates. The myths and legends which the initiates gave to the people depicted "the same experiences which came, as if in a living dream" during the three-day temple sleep. Ancient humanity's limited sense of self, of "I-ness," would not permit most people to undergo the Mystery rites; thus he received from the priesthood "pictorial representations of things which appeared to spiritual vision" during the initiation. Steiner cautioned that

> The pictures comprising the content of myth must not be confused with a merely symbolical or allegorical interpretation. The pictures comprising the content of a myth are not invented symbols for abstract truths, but real soul experiences of the initiate. He experiences the pictures with spiritual organs of perception as the normal man experiences the representations of material things with his eyes and ears.[12]

With Greek history, Steiner observed a gradual movement from a divinely-ordered society to an earthly civilization. No longer could the Greek feel related to the Cosmic All as had the oriental world; the Greek did not see or feel the spiritual world in the same way. To the contrary, he had a much less real perception of the spirit in nature, and felt that it was only possible to have mental pictures of the gods—of earthly-cosmic relations. Steiner traced the shift from the "mythopoeic"" to the conceptual, as Greek philosophers increasingly used concepts to describe events and processes previously expressed through pictures.[13]

### Greek Mysteries

In the great crossroads of peoples that was ancient Greece, two Mystery schools met. Steiner called these the Northern and Southern Mysteries, from the paths of migration their leaders took after the natural catastrophe known to most mythologies and religions as "the Great Deluge." The Northern mysteries such as the Persian-Zarathustrian, the Chaldean, and the ancient Druids of northern Europe, featured a "macrocosmic" approach, in which the aspirant sought the spirit behind the veil of the physical senses, the spirit in outer Nature.[14] The Southern mysteries, which included those of ancient India and Egypt, took a "microcosmic" attitude, a path of a mystical kind, in which the aspirant tried to sink more and more into his own inner being. In Greece, these two streams manifested as the inward-turning Dionysian cult and the macrocosmic Apollonian mysteries.[15]

The German philosopher Nietzsche felt this polarity as it revealed itself in Greek tragedy. Nietzsche pointed to the ordered, balanced Apollonian attitude and the sometimes frenzied ecstasy of the Dionysian. Rudolf Steiner spoke of the great influence of the Eleusinian initiates—and Aristotle clearly showed how by catharsis the tragic drama attempted to purify the feeling life of the people, just as the pupil of the Mysteries had to undergo a much more intense purification.[16]

For the initiate, the forces that create the material world also exist in his own soul, and it is these forces which also create the great myth-pictures of the uninitiated general populace. As the intellect developed, many men could see that these myths were of human creation, even if divinely inspired. But the initiate, who senses a higher reality closed to the intellect, awakens in his own soul the slumbering force of the Hidden God—hidden because he exists everywhere in nature, yet invisible to the senses or to the intellect. This God, out of pure Love, allows himself to be divided into the myriad objects of the natural world where he lies hidden, as if in an enchanted tomb. By gazing into his own soul, the initiate releases the same sleeping, hidden force that lies enchanted in nature. Thus he considers his soul to be the mother who can conceive a god, if fertilized by nature; but because God is hidden in nature, the birth seems to be virginal. Steiner wrote that

> The Father rests in concealment, the Son is born to man out of his own soul. Thus mystic cognition is a real event in the cosmic process. It is the birth of an offspring of God. It is an event as real as any other natural event, only on a higher level. This is the great secret of the mystic, that he himself creatively releases his divine offspring, but he also prepares himself beforehand to acknowledge this divine offspring created by himself. The non-mystic lacks the experience of the father of this offspring.[17]

Steiner discussed the pre-Socratic philosophers and showed how their thoughts usually are fully comprehensible only if one is familiar with Mystery wisdom. One of the least decadent Mystery centers, the great Temple of Artemis at Ephesus, preserved the best of oriental wisdom, and Steiner found that among others, Heraclitus, Pythagoras, Plato, Aristotle, and Alexander the Great were deeply familiar with the Ephesian teachings.[18] Heraclitus (535-475 B.C.) occupied himself with the typically Greek problem of finding the eternal among the ephemeral, transitory phenomena of the senses. He blamed most of humanity's troubles on his confusion of the transitory with the eternal, and he placed the true battleground of these opposites in the human soul, where the ever-changing lower human nature battled with and obscured his higher nature—his *daemon*, or individual spirit. Heraclitus expressed his philosophy aphoristically in poetic images. He urged men to cease to value the transitory, for it was just this clinging to the transitory with his cognition that constituted the "original fault" or "original sin" of mankind. By dissolving transitory thoughts in the fire of the spirit, humanity could regain his lost innocence and become reborn in the eternal, bringing to birth thy spirit within his own soul.[19]

An important step from the older Mysteries toward conceptual thought took place in the community established by Pythagoras in the sixth century B.C. Pythagoras (ca. 585-500 B.C.) wandered to a number of Mystery centers where he sought initiation, particularly to those of Ephesus and Chaldea. In his community, the aspirants underwent, in addition to the usual purifications, a rigorous training in mathematics, from which they thought the principles behind the natural world could be derived. The Pythagoreans found great significance in the fact that they could determine natural laws entirely through their own soul-activity—through their thinking—and that the observation of nature then revealed the working of the laws they had established in their own soul. It became apparent to them that humanity "must first find these harmonious patterns within himself if he wishes to behold them outside in the cosmos."[20] For Pythagoras, then, the meaning of the cosmos dawns in the human soul:

> The eternal pattern therefore lies hidden in the depths of the soul. Let us de-

> scend into the soul, and we shall find the eternal. God, the eternal cosmic harmony, is within the human soul. The soul is not confined to the physical body enclosed by man's skin. For in the soul are born the patterns according to which the worlds circle in space.[21]

We can see, according to Steiner, that Pythagoras considered such thinking, which awakened the eternal within the soul, as a communion with the eternal. The Pythagorean education attempted to foster an increasing spiritual communion through cognition; thus it was a philosophical initiation.

Another example of the emerging intellectual thinking used as a form of initiation occurred at Plato's Academy. Plato (427?-347 B.C.) intended his dialogues to be a form of initiation into a cognition not dependent on sense perception. He hoped that the thought-experience engendered by his works would supplant the temple-sleep experience. As Steiner wrote:

> Plato's world-conception aims to be a form of cognition which in its whole nature is religion. It brings cognition to the highest man can reach in his feelings. Plato allows cognition to be valid only when it completely satisfies man's feelings. Then it is not pictorial knowledge; it is the content of life. It is a higher man in man. The personality is but a higher image of this man in man. In man himself the superior, the archetypal man is born.[22]

In other words, insofar as was possible, the dialogues were to be "a literary form of the proceedings of the Mystery places."[23]

In the *Timaeus*, Plato described an Egyptian initiation. He asserted that the Father God lies spellbound in nature and could be found only by one who awakened the divine within himself. The Father created the body of the world by diffusing himself into it. "On this body of the world, the soul of the world is stretched in the form of a cross. This soul is the divine element in the world. It has met with death on the cross in order that the world may exist."[24] Plato regarded nature as the tomb of the divine, which could be resurrected only in the soul of the initiate through enhanced cognition. This wisdom, begotten in the maternal human soul, Plato called the "Son of God", and they considered it to be the essence of the world-intelligence, or the *Logos*.

In a somewhat similar description of an initiation, found in Plato's *Symposium*, Socrates recounts the revelation he received from the priestess Diotima, which he said awakened the divine force in his soul. Socrates characterizes the soul as the "mother of the divine" in which is born the "Son of God, Wisdom, the Logos."[25]

Rudolf Steiner used this perspective to discuss the myth of Dionysus, the son of Zeus and Semele, a mortal woman. A lightning bolt slayed Semele, and Zeus ripped Dionysus out of her belly, keeping the premature fetus in his own thigh until Dionysus was mature. But Hera, Zeus' wife, stirred up the giant Titans against the boy, and they ripped him to pieces. Pallas Athena rescued the still-beating heart and brought it to Zeus, who once again gave Dionysus life. Steiner saw Dionysus as the union of the divine and the temporal elements in the soul, wherein the divine causes the soul to experience a great longing for its true spiritual nature. Hera represents a consciousness that is jealous of the higher consciousness, and she stirs up the lower nature of humanity—the Titans—which dismember the child of God. This can be seen in the human being, Steiner wrote, as the fragmented, materialistic intellectuality, such as in scientific thinking. But Zeus, the higher wisdom, forges a new harmonizing force, the Logos: born of a mortal mother (the soul) and a God.[26] In the reborn Dionysus we have a picture of enhanced cognition.

Steiner emphasized that these pictures do not represent only soul processes, taking place entirely within the soul. Rather, the true spiritual reality is that through the myth the soul does not merely experience something within itself; it is completely disconnected from itself and participates in a cosmic process which in truth takes place outside itself and not within it.[27]

Steiner maintained that very definite laws govern the "unconsciously creative soul" and allow it to participate in the spiritual events which it clothes in mythological pictures. But the pictures are built according to these laws, so that when the initiate penetrates behind the pictures, the forces he observes are at one and the same time present in the myths, in his soul, and in the cosmos. Yet the spiritual forces and events themselves are totally *supersensible*, and the pictures are reminiscent of the material world; thus the myth pictures

> are not in themselves spiritual, but are merely an illustration of the spiritual. Whoever lives only in pictures lives in a dream; he lives in spiritual perception only when he has reached the point of experiencing the spiritual in the picture, just as in the material world one experiences the rose through the mental representation of the rose.[28]

Several conclusions may be drawn from this discussion. One is the closeness of the mythopoeic consciousness to our modern dream consciousness. While the emerging intellect was much better equipped to delve into the material world and to weave intricate philosophies, the danger was that it would destroy the divine element in cognition, while the old consciousness receded more and more into the chaotic world of the dream. At one time, clairvoyance provided a clear picture of spiritual beings. Later, this spiritual sight deteriorated into a dreamy feeling for spiritual forces and the mythological pictures became more of a tradition than a direct perception.

In spite of the efforts of those men like Plato and Heraclitus, the Greeks and Romans increasingly mistook the temporal for the eternal. The Hidden God seemed to become more hidden and, as we shall see, the Mysteries also slipped under the surface of history's flow. Steiner summed up the secularization of Greek culture by comparing Heraclitus' offering his work, *On Nature*, in the temple at Ephesus to Plato's statement, despite the grandeur of his own work, that "all the philosophy of his time was as nothing compared with the ancient wisdom received by the forefathers from the spiritual worlds themselves."[29]

Finally, Steiner pointed to the Herculean efforts of Aristotle, who attempted to supply a scientific, logical basis for Plato's Theory of Forms, as an abstraction, a reducing to concepts of the living ancient wisdom. Yet as a last flame of the ancient stream,

> something of the old wisdom still breathes in his works. In his concepts, in his ideas, however abstract, an echo can still be heard of the harmonies which resounded from the temple-sanctuaries and were in truth the inspiration not only of Greek wisdom but also of Greek art, of the whole folk-character.[30]

## Agriculture and Mythology

Most people know of the intimate relation between agriculture and mythology, and I devoted some attention to it in the previous chapter, particularly as agriculture was administered by the temples. Many authorities have discussed the numerous variations of the cereal-goddess myth, from Demeter and Persephone of the Greeks, Ishtar among the Babylonians, to Isis in Egypt and Ceres in Rome. For example, the Jungian psychologist, Erich Neumann, wrote a fascinating study of these world-mother myths called *The Great Mother*.[31] The point I want to emphasize here is Steiner's concept of the myth as a picture of real natural forces and their spiritual counterparts.

For the initiate, as we have seen, the same force that created the myth in his soul was present in outer nature and also in the spiritual world. To see behind the myth to a spiritual reality was to stand outside one's soul and participate in the cosmos, in cosmic and earthly processes, which at the same time also took place within the soul. While it would take many centuries for it to trickle down into agriculture to the level of the small farmer and gardener, the shift of consciousness from the old picture consciousness to intellectual thinking meant a slow and gradual loss of this *active participation* in the world order through cognition.

The intellect alone does not participate; instead, it stands aloof, merely observing and putting the percepts into order. The dream-picture of myths was torn asunder, but without the healing higher consciousness to forge a new reality. Steiner explained that

> Our modern highly developed intellect is, fundamentally, a late development of what, in the East, was dream-like clairvoyance. This dream-like clairvoyance has cast off its direct insight into the outside world and evolved into our inner logical order—into the great modern means of acquiring knowledge of nature.[32]

Yet until the end of the fourth epoch in 1413 A.D., agriculture and horticulture remained somewhat safe from the deleterious effects of the intellect. A mighty folk-wisdom preserved much of the old agricultural mystery-knowledge among the peasantry, but this all too easily degenerated into tradition or worst—superstition. With the onset of abstract, conceptual thinking, we begin to approach the time when enhanced cognition will once again become necessary. But before we can continue our journey, we must reckon with what Rudolf Steiner called the turning point in human evolution.

The historian Lynn White, Jr. wrote:

> The victory of Christianity over paganism was the greatest psychic revolution in the history of our culture. . . Our daily habits of action, for example, are dominated by an implicit faith in perpetual progress which was unknown either to Greco-Roman antiquity or to the Orient. It is rooted in, and is indefensible apart from, Judeo-Christian teleology. The fact that Communists share it merely helps to show what can be demonstrated on many other grounds: that Marxism, like Islam, is a Judeo-Christian heresy. We continue today to live, as we have lived for about 1700 years, very largely in a context of Christian axioms.[33]

Admittedly, this is a sweeping statement, but one with which, I think, Rudolf Steiner would have agreed. In order to discuss Steiner's ideas on scholasticism and the origins of modern science, and thus to lead into medieval agriculture, we must consider Steiner's views on Christ and Christianity.

## Christianity and the Turning Point

*Christianity is only at the beginning of its activity, and its real mission will be fulfilled when it is understood in its true spiritual form.*

*What Christianity bestows goes with us into all ages of time to come and will still be one of the essential impulses in humanity when religion, as we know it, is no longer in existence. Even when religion as such has been transcended, Christianity will remain. The fact that it was first of all a religion is connected with the evolutionary process of humanity. But Christianity as a world-view is greater than all religions.*[34]

*Rudolf Steiner*

From the discussion of Steiner's views on myths and the Mysteries, perhaps it will be apparent that he considered religion to be a stage in the evolution of consciousness, although an important one: a stage between the loss of the old clairvoyance and the development of an enhanced cognition and perception that will bring a new clairvoyance based on individual human freedom. Steiner compared the initiate to the modern botanist:

> As a botanist contemplated the growth of a plant to discover its laws, so the mystic wished to contemplate the creating spirit. The mystic was conscious that by seeking the truth contained in a myth, he was adding something to what was present in the consciousness of the people. It was clear to him that he was placing himself above this consciousness just as the botanist places himself above the growing plant.[35]

But the old Mysteries grew decadent, the intellect superseded clairvoyance, and men increasingly saw the human side of myth, not the divine.

The task of religion, then, is identical with the meaning of the Latin word from which it is derived—to "bind back." Religion arose from the knowledge revealed to the initiates who still possessed clairvoyance. It provides a body of revelation to which men can turn through faith until the time comes when they will have regained modern clairvoyance through their own efforts.[36] The age of faith was beginning; the ages of supersensible knowledge were rapidly ending. It was at this crucial time, shortly after Rome had changed forever to a dictatorship, that Jesus of Nazareth was born in Israel, a birth and short life that went virtually unnoticed in the world of classical antiquity.

Yet the Gospels of Luke and Matthew tell of two significant groups who did notice: the "Wise Men" from the Chaldean-Babylonian Mysteries and the simple shepherds. For Steiner, these Magi represented the knowledge of the old Mysteries, while the shepherds marked a new departure. Several spiritual streams met in the Incarnation of Christ. The northern and southern Mysteries—or macrocosmic and microcosmic—began to merge; thus the inner and outer paths became less differentiated. But perhaps the most significant aspect was the presence of the shepherds, because for the first time, non-initiates became cognizant of a knowledge previously attainable only through a Mystery center.[37] Through the shepherds, we have a picture of cognition attained by a new path: faith.

We have seen Steiner's description of the Mystery experience and its evolution into the Neo-Platonist concept of the "Son of God" coming to birth within the soul of humanity. This union with the spirit through cognition, whether attained through the highest philosophy, as with Pythagoras and Plato, or through the temple-sleep, was the concern of only a few people, a very small, elite group. For Steiner, one of the most important aspects of the life of Christ was the broadening of initiation beyond the temple and the spiritual communities such as the Pythagoreans, the Essenes, or the Therapeutae:

> Among the Essenes a whole community cultivated a life by which its members were able to attain this "union;" through the Christ event something—that is, the deeds of the Christ—was placed before the whole of humanity so that the "unionÆ became a matter of cognition for all mankind.[38]

Steiner considered the Christ to be the Logos, the Cosmic Word, the Son of God. The Neo-Platonists thought, as we have seen, that:

> The world has come forth from the invisible, inconceivable God. A direct image of this Godhead is the wisdom-filled harmony of the world, out of which material phenomena arise. This wisdom-filled harmony is the spiritual image of the Godhead. It is the divine Spirit diffused in the world; cosmic reason, the Logos, the offspring or Son of God.[39]

Through cognition, humanity can unite itself with the Logos, and in doing so link themselves with both the spiritual world and the sensory world; in other words, the human being becomes the link between the Spirit and matter. The Logos is resurrected spiritually in the soul through cognition, "if His creative word is understood and re-created in the soul."[40]

Rudolf Steiner declared that in the Baptism of Jesus at the river Jordan, this Logos, this mighty spiritual being, descended into the human body of Jesus and remained there for some two and one-half years, until the Crucifixion. If we recall Plato's image, drawn from the Mysteries, of the Macrocosm: God has stretched the soul of the world—the Logos—on the body of the world *in the form of a cross*, we see a picture which became in the life of Christ on earth, an earthly, physical *fact*. As the John Gospel says, the Word became flesh—was incarnated.

Yet, said Steiner, this cosmic spirit, whom we saw in the Indian epoch and the Persian epoch as the Sun-Being, and whom the Egyptians called Osiris, not only inhabited a human body, but also followed the human soul through the portal of death. The Christ then arose from the Dead, from Hades. This death and resurrection Steiner termed the Mystery of Golgotha. It formed, he said, the culmination of the ancient Mysteries, and since it took place in world history, since the Word had become flesh, it opened the way for all men to share in it. Supersensible cognition became even more dim in the Dark Age, yet all men could attain to knowledge of the Mystery of Golgotha through faith. Faith, if it is enhanced through prayer and meditation, can become a bridge to a renewed, Christ-filled spiritual vision.

The remainder of the fourth epoch saw faith, and its companion, religious dogma, assume an increasing dominance in religious affairs. Philosophy rose to great heights during the ensuing seven centuries, culminating in scholasticism and its greatest exponent, Thomas Aquinas. At the same time, philosophers saw cognition as increasingly limited, suitable for dealing with ideas concerning the material world, but incapable of directly comprehending spiritual reality. Supersensible cognition became a matter of faith by the Middle Ages, yet through faith all men could attain the truth to which only the few achieved through cognition in the old Mysteries. The Mystery of Golgotha "brought the content of the Mysteries out of the darkness of the temple into the clear light of day," where it was to be perpetuated largely in the form of faith and dogma.[41]

### The Christ and Human Individuality

Steiner described Moses as an initiate who prepared humanity for the coming of the "I am", or ego consciousness. We have seen Steiner's concept that the succession of civilizations is not meaningless, but reflects different stages of consciousness, and each people has a particular mission to fulfill. Moses mastered the deepest secrets of the Egyptian mysteries and restated them for the Hebrews.[42] It fell to these people, Steiner explained, to prepare themselves to receive the incarnation of the Logos in their midst, and he outlined the progress of the forty-two generations that lived between Abraham and Jesus which were necessary to prepare a physical vehicle for the Christ. Yet even the Egyptians felt as early as 2200 B.C. that they were losing their supersensible perception. In their mysteries after the time of Moses, the Egyptians could no longer find union with the Cosmic Word, and any contact with the spiritual world could only be reached through decadent magical practices. By the time of Christ, the Hebrew people had come to the end of their line of mighty prophets; the voice of God—the *Bath Kol* as they called it—no longer seemed to speak through them. Steiner showed the development from Moses, who gave the Hebrews the Law to govern their behavior—a clear indication that their "I" was little present in them—to Jeremiah, who told them that Jahweh would "put my law within them and write it on their hearts." And it was Elijah who experienced God within him and not in the outer world: neither outside in the spiritual, nor outside in the perceptual world, but *within.*

Steiner contrasted Christ's two commandments to those of Moses: Christ gave only two—"thou shalt love the Lord thy God with all thy heart, and thou shalt love thy neighbor as thyself"—and these are incapable of being enforced. Rather, they depend on *love offered freely,* and this can only be done by a morally free "I".[43]

The possibility for such a free moral act, Steiner said, became a possibility for *all* men only with the Christ. St. Paul preached that the Law had been a "schoolmaster," but that Christ brought the possibility of freedom, releasing humanity from the "bondage to the Law."[44]

The Mystery of Golgotha gave all people the opportunity to accept the Christ into their own inner being, into their lower self. St. Paul's saying "Not I, but Christ in me," thus explains the means by which men can obey the commandment that Paul regarded as the fulfillment of the Law , to love one's neighbor as oneself: because the Christ is the spiritual being who becomes the higher self in humanity when He is received into the soul. As we have seen, this "re-birth" in the human soul, after the Mystery of Golgotha, could take place in all men, through either cognition or faith. Steiner asserted that Christ referred to himself as the Higher Self (the 'I am') many times, but that these passages are translated wrongly, since the translator was unaware of their significance. For example, such familiar sayings as "I am the Good Shepherd", etc. should read "The I am is the Bread of Life, The I am is the Good Shepherd, The I am is the Light of the World, The I am is the True Vine." The "I am" refers both to the Being of Christ and to the Higher Self of humanity.[45]

Steiner also pointed out that Christ spoke of the divisive effects of the lower, personal "I" when Christ said: "Think not I am come to send peace on earth, I came not to send peace but a sword." (Matthew 10: 34-35)

> Suppose ye that I am come to give peace on earth? Nay, but rather division; for henceforth there shall be five in one house divided, three against two, and two against three. The father shall be divided against the son. . . (Luke 12:51-53)

Such egotistic behavior invariably results from the "un-Christened" I, or ego.[46]

But it is not necessary for men to belong to any Christian sect for "re-birth" to occur. In the next chapter we will deal with both the esoteric Christian development and the path of faith. Here we must emphasize Steiner's assertation that the Christ was the spiritual being who had concerned himself with humanity throughout human evolution, and who had united himself with both the human soul and with the earth itself in the "Ascension." Christianity as a religion occupies only a portion of this evolutionary course.

Several points should be stressed, since Steiner returned to them many times.

The Mystery of Golgotha occurred at the time when the individual personality was becoming very strong. This had great import for Rome and the subsequent development of Christianity as a religion, as did the other characteristic of the fourth epoch: the growing power of the intellect. For that story, we must move toward the Roman Empire and the Middle Ages.

I can only emphasize that Rudolf Steiner gave the Mystery of Golgotha a central place in all his teachings, particularly the development of consciousness. I must refer the interested reader to the appendix for a list of some of Steiner's most comprehensive books on Christianity.[47]

**Notes for Chapter 4**

1. Quoted in Rudolf Steiner, *Christianity as Mystical Fact and Occult Mysteries of Antiquity* (Blauvelt, New York: Rudolf Steiner Publications, 1961) 79, 55, 53. Cited hereafter as *Christianity as Mystical Fact.* This edition contains an excellent biographical sketch of Steiner by Alfred Heidenreich, and helpful notes by Paul M. Allen.
2. Ibid.
3. Ibid., 95.
4. Ibid., 49-50.
5. Quoted in D. J. van Bemmelen, *Zarathustra*, I (Zeist: Uitgeverij Vrij Geestesleven, 1968) 41.
6. At least the outline of the Mystery experience is well known to historians. For example: Luigi Pareti, Paolo Brezzi and Luciano Petech, *The Ancient World*, 1-2 (London: George Unwin and Allen, 1965) 236:

> Initiates had to undergo *symbolically* the sequence of death, journey to the underworld, and resurrection into a new life. In the second phase of this sequence, they attained to divine knowledge of hidden truths, being shown sacred representations of the divine myth, seeing and touching sacred objects, and partaking of sacrificial food, all under the influence of intoxicating drink or during a trance. (my emphasis).

7. Rudolf Steiner maintained that this experience was far more than symbolic. See note 12. Rudolf Steiner, *Christianity as Mystical Fact*, 49-51.
8. Ibid., 52-53,55.
9. Ibid., 65, 204.
10. Rudolf Steiner, *The Tension Between East and West* (London: Hodder and Stoughton, 1963) 120-121.
11. Rudolf Steiner, *Turning Points in Spiritual History* (London: 1934) 280; Steiner, "European Mysteries and their Initiates", *Anthroposophical Quarterly* (Spring, 1964) 170, 171.
12. Rudolf Steiner, *Turning Points*, 281; *Christianity as Mystical Fact*, 113. Compare Steiner's statement to Pareti's in note 6.
13. Rudolf Steiner, *World History in the Light of Anthroposophy* (London: 1950)79-84.
14. For the Druids, see Steiner. *The Evolution of Consciousness* (London: 1966); and Steiner, *Mystery, Knowledge and Mystery Centres* (London: 1973, 2nd ed.) especially lectures 7-9.
15. D. J. van Bemmelen, *Zarathustra*, I, 42-50; Steiner, *Turning Points*, 54-55, 58-59.
16. Rudolf Steiner, *Nietzsche: A Fighter for Freedom* (Blauvelt: Rudolf Steiner Publications, 1971).
17. Steiner, *Christianity as Mystical Fact*, 67.
18. Steiner, *World History*, 85.
19. Steiner, *Christianity as Mystical Fact*, 69-73.
20. Ibid., 82.
21. Ibid.

22. Ibid., 99.
23. Ibid., 93.
24. Plato quoted in Ibid., 95.
25. Ibid., 101.
26. Ibid., 102-103.
27. Ibid., 103.
28. Ibid., 114.
29. Rudolf Steiner, *Occult History* (London: Anthroposophical Publishing Co., 1957) 101.
30. Ibid., 102.
31. Erich Neumann, *The Great Mother* (Princeton: Bollingen Press, 1965); see also Pareti, Brezzi and Petech, *The Ancient World*, II, 119.
32. Steiner, The Tension Between East and West, 153.
33. Lynn White, Jr., "The Historical Roots of Our Ecological Crisis", *Science*, 155 (10 March, 1967) 1205.
34.Steiner, *Christianity as Mystical Fact*, frontispiece.
35. Ibid., 105-l06.
36. Stewart C. Easton, *Man and World in the Light of Anthroposophy* (New York, 1975) 175. Cited hereafter as *Man and World.*
37. For a good discussion of this see F. W. Zeylmans van Emmichoven, *The Reality in Which We Live* (East Grinstead, Sussex: New Knowledge Books, 1967).
38. Steiner, *Christianity as Mystical Fact*, 177. See also René Querido and Hilmar Moore, *Behold, I Make All Things New: Toward a World-Pentecost* (Fair Oaks: Rudolf Steiner College Publications, 1990).
39. Ibid., 189.
40. Ibid., 190.
41. Ibid., 201.
42. Steiner, *Turning Points*, 183, 185.
43. Easton, *Man and World*, 41-42, 51-53; Rudolf Steiner, *Mysteries of the East and of Christianity* (London: 1972, 2nd ed. rev.) 50-51.
44. Easton, *Man and World*, 52.
45. Ibid.. 191, quoted from Emil Bock *The Three Years* (London: Christian Community Press, 1955; 108-110.
46. Easton, *Man and World*, 191.
47. A helpful selection of books for a deeper understanding of Steiner's picture of the Christ are: Zeylmans van Emmichoven, *The Reality in Which We Live* (East Grinstead: New Knowledge Books, 1967), the best single starting point for the anthroposophical view of esoteric Christianity; Emil Bock, *The Three Years: The Life of Christ Between Baptism and Ascension* (Edinburgh: Floris Books, 1955); Friedrich Benesch, *Easter* (Edinburgh: Floris Books, 1982); Benesch, *Whitsun* (Edinburgh: Floris Books, 1982); Benesch, *Ascension* (Edinburgh: Floris Books, 1983); and René Querido and Hilmar Moore, *Behold, I Make All Things New* (Fair Oaks: Rudolf Steiner College Publications, 1990).

## CHAPTER FIVE

## THROUGH THE MIDDLE AGES TO MODERN TIMES

As we move from Rome through the "dark ages" to the Middle Ages, we move from the climate and light soil of the Mediterranean to the heavy soil and dense forests of northern and western Europe, and we meet the various Germanic and Celtic tribes who gradually pushed all the way south to the shores of the Mediterranean itself. Our story must consider these vigorous people and the civilization they forged out of foundations left by the classical world. The effect on the evolution of consciousness and agriculture was quite far reaching.

Rudolf Steiner described the European "barbarians" as a people who preserved the ancient dreamlike clairvoyance to a high degree until fairly recently. Many of them, he wrote

> . . . were acquainted with the spiritual world through their own experience and were able to communicate what took place in that world to their fellow men. A treasure house of narrative about spiritual beings and spiritual events was built up. The treasures of folk fairy tales and myths arose originally from such spiritual experiences.[1]

Others had developed faculties of perceiving the sensory-physical world that corresponded closely to clairvoyance. The Mysteries of these people sought to inculcate the perception of the spiritual beings who move behind the forces of nature, and their mythologies contained the remnants of this knowledge.

Steiner portrayed these Europeans as a people who retained both actual clairvoyance and also feelings and sensations that yearned strongly for the spirit, yet who also turned their attention more and more toward the world of the senses and its control. Thus began a division in the European soul: on the one hand lay the mystics like Eckhardt and Tauler, whose mystical feelings were permeated by Christianity; on the other hand was the increasing mastery of the material world, the rise of modern science and technology. For Steiner, the progress of outer culture that began in the Middle Ages and accelerated rapidly in the fifth epoch resulted from just this separation and the strong tendency toward physical life.[2]

From their many Mystery centers emanated myths which prepared the Europeans for the coming of Christianity, such as the well-known myths of the Celts and of the death of the Norse god Baldur and the ensuing "Twilight of the Gods".[3] Even these Mystery places were turned toward the world of nature, toward a spiritual understanding of nature. The Roman Catholic church waged a long and bloody war against them and against the old clairvoyance, the course of which is, in itself, a fascinating study in the light of Steiner's research. What concerns us here, however, is the influence the Mysteries and the soul character of the Europeans had on the development of European agriculture.

The plentiful summer rains and heavy soils of Northern Europe demanded different techniques than those used in the Mediterranean lands. Farmers responded with a number of innovations which, beginning in the fifth or sixth century and extending over the next several centuries, revolutionized agriculture. The development of the heavy wheeled plow allowed the European farmer to exploit the rich alluvial bottom soils which are more productive than the hillside soils. By the ninth century, the nailed horseshoe and improved harness led to a much greater use of the horse in farming. One historian showed that the old yoke-harness permitted a team of horses to pull about 1,000 pounds, whereas with the new collar-harness a team could pull four or five times as much.[4] The demand of armored knights for larger horses provided better breeding for plow horses as well. Given the much smaller size of cattle in the early Middle Ages, the change to horse-power was significant.[5]

Along with these developments came the extensive use of the three-field system, called "the greatest agricultural novelty of the Middle Ages in Western Europe."[6] By dividing the arable land into three parts of winter grains, summer crops, and fallow, the total acreage could be increased by one-third; by the twelfth century, this was increased to fifty percent. Labor thus was fifty percent more productive and gave an impetus to the great movement toward land reclamation: swamp-draining, forest removal, dike-building. The new use of legumes added a missing protein factor to that of the winter grains, improving nutrition dramatically.

So we can see that by the tenth century, these innovations had increased productivity to the point where the focus of power moved from the classical south to the north of Europe. As Lynn White, Jr. expressed it:

> The agricultural revolution of the early Middle Ages was limited to the northern plains where the heavy plow was appropriate to the rich soils, where the summer rains permitted a large spring planting, and where the oats of the summer crop supported the horses to pull the heavy plow. It was on those plains that the distinctive features of both the late medieval and of the modern worlds developed. The increased returns from the labor of the northern peasant raised his standard of living and consequently his ability to buy manufactured goods. It provided surplus food which, from the tenth century on, permitted rapid urbanization. In the new cities there arose a class of skilled artisans and merchants, the burghers who speedily got control of their communities and created a novel and characteristic way of life, democratic capitalism. And in this new environment germinated the dominant feature of the modern world: power technology.[7]

Lynn White, Jr. said of these agricultural developments, especially the heavy plow:

> Formerly man had been part of nature. Nowhere else in the world did farmers develop an analogous agricultural implement. Is it coincidence that modern agricultural technology, with its ruthlessness toward nature, has so largely been produced by descendants of these peasants of Northern Europe?[8]

We have seen in the preceding chapter Lynn White's view that the beliefs of

Christianity led thinking toward the exploitation of nature, and that Steiner's picture is somewhat more complex. Along with the development of Christian philosophy, which augmented the work of Aristotle and forged a strengthened cognition that, in the philosophy of Albertus Magnus and Aquinas achieved a power not surpassed to this day, grew a significant movement of esoteric Christianity, of Christian Mysteries. We must consider both the cognition that led humanity to exploit nature with a thoroughly materialized thinking, and the attempts to lead this cognition once again to spiritual knowledge.

I believe that such an inquiry might cast some light on the thorny historical question of how certain inventions came about, just as Steiner's concept of the Mystery centers affords a helpful view of the origins of agriculture in the first and second epochs. Rudolf Steiner likened the flow of history to a mighty river:

> We cannot simply derive its features at a given point from what lies a little farther upstream, but must realize that in its depths there operate all kinds of forces that may come to the surface at any point, and may throw up waves which are not determined by those that went before.[9]

He maintained that simple causal explanations for historical events and forces could never do justice to reality. Thus the following description attempts to show the interweaving of various streams in the history of agriculture.

### From the Druids to the Cistercians and Chartres

Steiner told of the far reaching work of Aristotle, for most of which we have only notes of Aristotle's lectures as a remnant. His pupil Theophrastus, whom we saw as an early agricultural writer, realized that the development of Western peoples precluded an understanding of Aristotle's natural scientific works. These, thought Theophrastus, could only be utilized in the East, where some last remains of the knowledge of earthly-cosmic correspondences still existed. Much of this work another Aristotelian pupil, Alexander the Great, took to the East in his conquests and subsequent establishment of academies and libraries. Theophrastus sent Aristotle's logic and philosophy to the West, where we saw it form the foundation for the scholasticism of Albertus and Thomas Aquinas.[10]

But medieval philosophy was more than abstract thinking. Not only did it pioneer an increase in the power of thinking (of great importance for the future; the content is not), but there still breathed something of the spirit in it. We can recall Steiner's assertion that there existed in Aristotle's philosophy, although in conceptual form, the living knowledge of the Mysteries. Even until the nineteenth century, Steiner said, this aspect of Aristotle survived, for his work was used as a basis for meditation, as a meditative exercise, giving to students training and discipline in cognition.

Here we see a development of importance for education, for the understanding of which I must refer the reader to Steiner's concept of three-fold man and the bodily basis for psychological functions in the chapter on Steiner's life and work. In Greece, the teacher was the gymnast, and he taught the whole human being—head, heart, and hands—for the gymnast believed that: "From out of the whole human body in movement—for the Gods themselves work in the bodily movements of man—something is born that then comes forth and shows itself as human understanding."[11]

The Romans already took something away from this wholeness: the gymnast gave way to the rhetorician. Nevertheless, speaking involves the whole being too—the heart, lungs and diaphragm—even if not as intensely as gymnastic movements. After the Middle Ages, however, the rhetorician no longer taught as the predominant educator, and the study of Aristotle no longer was a meditative exercise. The new teacher, called by Steiner the "doctor" or the "professor," only cared about the head and its *thoughts*. No longer did educators inculcate abilities gained through exercises of self development; rather, all that mattered was the accumulation of abstract knowledge—the situation with most modern education.

But we are outrunning our story, and we must now account for the other Aristotelian stream. For alongside the change to abstract knowledge, there developed a powerful body of folk wisdom which somewhat softened the trend to intellectualism. This stream entered the West from several directions: from the European mysteries and their initiates, myths, and fairy tales; from the natural scientific teachings of Aristotle, which arrived in the form of certain Gnostic and alchemical doctrines from the Arab world; and from the lingering traces of clairvoyance and the deep feeling for the spirit in nature of the Europeans. This stream was very active in the rise of Christian esotericism, and we should recall that Aquinas' teacher, Albertus, and Aquinas himself were thoroughly knowledgeable in the doctrines of alchemy.[12]

Thus a deep knowledge of the spirit in nature and in the human soul coexisted with the growing intellectual world conception, and even some philosophers retained this knowledge. For example, Steiner said that Albertus Magnus knew of the existence of spiritual beings behind physical appearances. Lynn White, Jr. wrote that "this view of nature was essentially artistic rather than scientific."[13] While only a small number of people were ever initiated into the esoteric teachings of Christianity, as opposed to those who followed the path of prayer and devotion, we can see that nevertheless, such knowledge flowed into Europe from many sources. They met and reached their highest artistic and philosophical expression at Chartres cathedral, in the School there, and in the agricultural practices of the Cistercian order.[14]

Chartres has been a holy place for several thousands of years, far back into the days of the Druids.[15] These ancient priests possessed a most profound knowledge of cosmic-earthly correspondences and they sought places where the earthly minerals were especially propitious for receiving cosmic influences. They particularly valued places rich in limestone, upon which they built their granite circles, the most famous of which is at Stonehenge in England. In his agricultural lectures, Steiner pointed out how the calcium of limestone mediates the cosmic forces of reproduction while the silica of granite mediates the nutritive forces. This is a basic tenet of his biodynamic approach. Chartres is a granite building erected on a granite hill, surrounded by a limestone plain. It is the tallest point for many miles around. René Querido and Louis Charpentier assert that members of the Cistercian order were fully conversant with these correspondences.[16]

If we hypothesize that indeed the Cistercians were deeply knowledgeable of such matters, we have an important clue as to the development of medieval agriculture, for the Cistercians are recognized widely as its pioneers. They made significant advances, for example in the use of limestone as fertilizer, called the greatest medieval agricultural innovation, and they helped greatly to expand the cultivated lands of Europe. They were at the vanguard of managerial and organizational innovations, and were especially known for their doctrine that we should help the Christ to redeem the earth itself. To this end they accepted as gifts lands which were unarable, and through their skills at clearing, draining, irrigation and fertilization, the Cistercians turned many hundreds of wild and useless acres into beautiful farms and gardens. Their influence on European agriculture was unparalleled and by 1152, they had 328 estates all over the continent, from Yorkshire to Eastern Europe.[17]

The Cistercians also struck a blow against serfdom, and recruited numerous lay brothers to work their farms, which were also notable for their large size. Henri Pirenne found that "Nothing could have been more different from the demesnes of the old manorial estates than these fine Cistercian farms, with their centralized administration, compact form, and rational exploitation." In addition, in the colonization of wastelands, the free status of the lay brothers had a marked effect on the rise of personal liberty. Also the preeminent sheep-farmers of England, their stock-raising added to the available stock of manure for fertilizing, and they helped to pioneer the three field rotation.[18]

Chartres itself is an edifice of great beauty and totally imbued with esoteric knowledge. It is as if the ancient mystery wisdom were recast in Christian form, and the statuary and carvings reveal the most detailed astronomical and historical data, all from the viewpoint of Christian esotericism. Steiner held that the Chartres masters knew that this knowledge of cosmic-earthly correspondences, and its use in agriculture and for healing, would have to be lost for many years. While some of the old knowledge survived in Cistercian abbeys into the twentieth century, the overwhelming majority was forgotten.[19] It seems likely, then, that the Cistercians were inspired by Christian mystery wisdom in their mission to heal both the earth and the human soul, and I believe it to be a reasonable hypothesis that the medieval "agricultural revolution" derived many of its innovations not simply from inventive peasants, but from the wisdom of initiates. It would carry us too far afield to consider also the Knights Templars, an esoteric order whose task it was to bring a Christianized social and economic order to Europe, their influence on cathedral-building, and their connection to the Chartres school. Nevertheless they, too, give evidence of a strong esoterically-inspired attempt to bring spiritual knowledge to secular life. For a number of years they were the leading bankers of Europe, and through their agents in the Holy Land and Asia Minor, they brought much of Aristotelian science to the continent.[20]

Thus we have come full-circle in our consideration of medieval agriculture. We have seen the influence of the Northern European Mysteries on the view of nature, and the influx of Aristotelian mystery-wisdom, and how these two streams united to supply a fertile foundation of folk-wisdom, which provided much insight into the spiritual basis of natural processes. The logical-philosophical Aristotelian stream, while it led directly into abstract modern thinking, also had a role as a powerful tool for self-development, no longer relegated to secret Mystery places. Alchemy, both in the folk-wisdom, and in the philosophical and scientific work of Albertus and others, not only laid the framework for much of the later scientific endeavors of the Renaissance, but preserved a knowledge of the spirit in nature and in the soul. Albertus Magnus, for example, wrote a very good study of plant nutrition and physiology called *De Vegetilibus*. His imaginative, pictorial approach can be seen in his calling the soil the plant's "stomach, which predigests food for the plant to draw into its roots."[21] So the picture that emerges from our study is of a society struggling on the one hand away from control by the Church, trying to evolve workable institutions; and on the other hand, a civilization permeated by various streams of esoteric Mystery wisdom. Agricultural education, then, appears to have been largely at the hands of the Cistercians with their deep knowledge, or carried still by the traditions of the folk-wisdom.

Stewart Easton points out that by the fourteenth and fifteenth centuries, however, this civilization was in evident and rapid decay. The Great Schism, the Hundred Years War, the Black Death and other signs of collapse abounded. Clearly a new age was beginning, an age that started with the revival of classical learning in Italy. We now turn from the medieval culmination of the fourth epoch (747 B.C. to 1413 A.D.) to modern times, the fifth cultural epoch.[22]

### A Brief Introduction to the Fifth Epoch

Perhaps the best way to introduce the basic soul mood of the fifth epoch would be to compare it to the previous one. In the fourth epoch, we saw the development of the intellect, of conceptual thinking, and by 1413 A.D., many people could think for themselves. Yet the intellect had not become the cold, dessicated instrument of today; rather, it worked together with the heart and a fertile imagination. If Aristotle, Theophrastus, and Albertus can be criticized for excessive generalization, nevertheless a certain intuition worked within them, and one senses a feeling of *understanding*, of a deep wisdom, in their writings. For example, unlike a modern scientist, Aristotle was not satisfied to describe the various parts of a thing and how they worked together; he also sought for its purpose and place in the world-order, what he called the final cause, and he was convinced that such purpose existed in the universe. Speaking of this intuitive intellect, Stewart Easton wrote that ". . . it seems to me that Steiner was indeed suggesting that these men, though *knowing* comparatively less than we, yet may have *understood* more, difficult though it may be for us to accept this possibility."[23]

The dawning of the fifth epoch brought a trend toward increased objectivity. The self could withdraw completely from the subject being studied. By comparing the modern philosophical problem of how the "subjective" mind can know the "objective"
world, to the earlier feeling that humanity and world were an inseparable whole, we can achieve an idea of the new epoch. As the epoch progressed, more phenomena came to be viewed "objectively," until even the mind itself was analyzed; humanity thus could take a somewhat objective view of itself. The old connections and correspondences were severed and forgotten. Rudolf Steiner called this dawning objectivity the "spectator" consciousness, and the familiar story of Galileo's discovery of the law of the
pendulum affords a good illustration of its modus operandi.

Sitting in church, Galileo observed a hanging lamp as it swung, probably due to a breeze. In order to measure the time-span of each swing, he used the beat of his pulse. Galileo, then, arrived at one of the basic laws of mechanics while attending Mass. As Ernst Lehrs wrote:

> Now, a faithful Christian who allows himself to be touched in his soul, however slightly, by what he believes to occur at the altar during the Holy Act, cannot but experience some change in the rhythm of his heart-beat. In order to use his heart as a chronometer, Galileo had to remain entirely unaffected, at best for a few moments, by the event, holy though it was for his belief. Thus we may imagine him sitting in the midst of the pious crowd, his gaze fixed on the swinging lamp, the finger of one of his hands on the pulse of his other wrist, while he carefully counts the beats—the picture of the perfect onlooker![24]

The origins of the scientific world-view have been discussed by Rudolf Steiner and several of his followers, most thoroughly by Ernst Lehrs in his *Man or Matter* and Owen Barfield's *Saving the Appearances*.[25] Both these men, and many of Steiner's lectures, delineated the great significance of the "onlooker" or "spectator" consciousness. While a full synopsis of their work exceeds the limits of this study, several points must be made.

In the fifth epoch, Europeans gained increasing mastery over the physical world. But just as at the beginning of the fourth epoch, we detected mythological pictures in the work of the early philosophers, a "throwback" to an earlier time, so at the start of the fifth epoch we see a great revival of classical culture. The Renaissance, while it heralds a new stage of consciousness, proceeded in the shoes of ancient Greece and Rome. In addition to the influence of classical culture on art and science, we can detect the strong influence of the ancient Mysteries, too. Rudolf Steiner spoke many times of this fact, and recent scholars are providing verification of his work as more records come to light. The historian Frances Yates, in two excellent books, *Giordano Bruno and the Hermetic Tradition* and *The Rosicrucian Enlightenment,*[26] while she does not take into account Steiner's ideas, and in fact her view does not agree with Steiner's conception of Mystery-wisdom, Dr. Yates nevertheless affords a fascinating view into the new abilities of the spectator-consciousness coming to terms with the old world-view.

The point is that the view of the human soul as a meeting ground of earthly-cosmic correspondences took quite a long time to die. Rudolf Steiner contended that a determined group of men consciously strove to keep such ideas alive, because they realized that the new faculties of detached observation, added to the powerful thinking developed in medieval scholasticism, could result in tremendous damage if focused entirely on the physical-sensory world. Writers such as Frances Yates now see the importance of the hermetic-alchemical and Rosicrucian traditions in the birth of modern science; they fail to see the stream of Mystery-knowledge from which these traditions arose. For Steiner, they were a direct continuation of the path of supersensible wisdom we have followed throughout this study, and out of which the best aspects of human culture have come. For most historians of science and philosophy, these men, despite their brilliance, were ensnared in a hopelessly outdated, nearly unintelligible world-view.

### The Birth of Modern Science and Mystery-Wisdom

The modern scientific world-view came into being over the past four hundred years, beginning with the work of Galileo, Kepler and Copernicus. It was born amid the great cultural upheavals of the Reformation, the Counter-Reformation, and the devastating Thirty Years War. During this period of religious strife and hatred, the pioneer scientists laid the foundations which came to full fruition in the nineteenth and twentieth centuries. Behind the scenes, more and more underground, the Mystery-wisdom strived to maintain a knowledge of humanity's spiritual origins and of the correspondences between earthly and cosmic phenomena. These men and their work, of whom Steiner gave a fairly complete picture, found themselves caught between fear, persecution, and war on the one hand, and the increasing materialism of the spectator-mind on the other. The secrecy that descended over their work like a shroud originated from these twin pillars of opposition. The historical records overflow with reports of massive book and manuscript burning during the fifteenth and sixteenth centuries.

In *Mysticism at the Dawn of the Modern Age*, Rudolf Steiner described the lives and work of eleven men who form a bridge between the medieval world-conception and modern science.[27] In addition to their own keen spiritual striving, most of these men studied alchemy very seriously, particularly the work of Albertus Magnus. Among them are Boehme, Johannes Tauler, Agrippa of Nettesheim, and Nicholas Cusanus. Especially relevant for our study of agriculture is the controversial Paracelsus, who, Steiner said, made a deep study of nature and learned all he could from the surviving folk-wisdom. His abrasive personality and the public derision he cast upon the tradition-bound physicians of the age, who referred to classical texts for every diagnosis, and upon the pharmacists, who sold adulterated and even worthless drugs, eventually cost him his life. The fact that his perceptions afforded him much greater success than his hapless competitors did not help his cause; nor did the fact that he often charged no fee! The great store of knowledge of nature he found among the country folk provides evidence of how powerful this stream of nature-wisdom remained. Also relevant for this study is the fact that many of his ideas presage the later, and much better known work of Samuel Hahnemann with the efficacy of highly diluted substances in medicine, which Hahnemann called homeopathic medicine. Several writers have said that the biodynamic preparations work in a similar manner as the homeopathic medicines.

Notable figures such as Paracelsus, Giordano Bruno, Dr. John Dee and Dr. Robert Fludd faced increasing opposition. Perhaps the most effective of those groups who battled to preserve spiritual knowledge from the powerful opposition it met on every side were the Rosicrucians. Paradoxically, it is this brotherhood of which external historical sources provide the least information. According to Frances A. Yates:

> Since the object of a secret society is to keep itself secret, it is not easy to uncover the secrets of the Rosicrucians; even the manifestoes explicitly concerned with them are concluded in veiled and mysterious language, and nothing is known of their organization, or even whether they had an organization or were actually formed into a sect.[28]

Rudolf Steiner considered the Rosicrucians to be the most significant esoteric movement in the fifth epoch.

Frances A. Yates concluded that Rosicrucian ideas had a tremendous influence, and in fact form the connecting link between the Renaissance hermeticists like Giordano Bruno and the early scientific revolution. She wrote that:

> The great mathematical and scientific thinkers of the seventeenth century have at the back of their minds Renaissance traditions of esoteric thinking, of mystical continuity from Hebraic or "Egyptian" wisdom, of that conflation of Moses with Hermes Trismegistus which fascinated the Renaissance.[29]

Dr. Yates cites particularly the influence of John Dee, who inspired the Elizabethan technical advance, the poets who gathered around Sir Philip Sydney, and many figures in Germany, such as Amos Comenius. Yates asserts that the Rosicrucians influenced many leading scientists, including Francis Bacon, Leibnitz, Johannes Kepler, Robert Boyle, and even Isaac Newton. She found in the Rosicrucian movement a union of religious and scientific vision. Many of those most closely connected with the movement also were influential in founding the British Royal Society, which was modeled along lines suggested in Rosicrucian literature. Robert Boyle's tutor was Peter Sthael of Strasbourg, a noted Rosicrucian, who also taught the architect Christopher Wren and John Locke. Sthael was for many years head of the Royal Society's laboratory.[30] The Rosicrucians in England and on the continent knew that major scientific discoveries were imminent, and they hoped through their religious-scientific knowledge to provide a basis to heal the terrible strife that raged over Europe. Of significance for our study of agriculture is Yates' finding that the Rosicrucians endeavored to establish many small local societies for disseminating their views.

Rudolf Steiner's descriptions of the Rosicrucians agree with much of Yates's work. Her books are works of superb scholarship. Steiner, however, did not limit himself to external sources; the picture he gave goes far beyond that of Dr. Yates. He asserted that the founder of the Rosicrucians, Christian Rosenkreutz (1378-1484), regarded by Yates and other scholars as only a legend, actually lived and worked in Europe. He traveled widely through Asia Minor and the Middle East and perhaps to Ethiopia and India, seeking the remaining sources of esoteric knowledge Upon his return, he gathered a circle of pupils and began to teach. Steiner said that Christian Rosenkreutz was one of the great teachers of humanity, and that he restated the ancient Mystery wisdom in a form suitable for modern people. The influence of his teaching, Steiner held, predates his birth and extends far past his earthly life. We shall see what evidence in the external life of Europe might corroborate Steiner's assertions based on his supersensible research.[31]

The full scope of Steiner's ideas on the Rosicrucians extends far beyond the limits of this study. Steiner's work shows, I believe, that the ideas and methods of the Rosicrucians form the link between the Mysteries of the Middle Ages, which culminated with the Chartres School, and the Mystery-wisdom of modern times. Many of the same characteristics of the work of the Chartres school, of the Cistercians in agriculture and the Knights Templar in socio-economics, appear in the Rosicrucians' work. Times had changed greatly, however, and so their methods were different; but the goals were similar—to provide a spiritual basis for life in the world. Steiner saw the Rosicrucian impulse to be the preservation of spiritual knowledge in the face of the impending scientific revolution; not to oppose the age of natural science, but to preserve knowledge of the spirit from the dangers of excessive materialism.

These men sought to engender a direct perception of the spirit in nature and in human nature, and the meeting of spirit and nature in the human soul. As Steiner wrote of the task of the Mysteries after the life of Christ:

> Previously this wisdom existed exclusively in order to enable the human being to bring himself into a soul state that allowed him to behold the kingdom of the Sun Spirit outside of earthly evolution. Now Mystery wisdom was allotted the task of making the human being capable of recognizing the Christ Who had become man, and from this center of all wisdom to understand the natural and spiritual world.[32]

To the increasing ability of the "onlooker" consciousness to delve into the material world, the Rosicrucians wanted to add the knowledge of the spirit and its workings, to heal the widening rift between the "subjective" inner life and the "objective" world of nature, between religion and science, Catholic and Protestant.

This break between the medieval and modern world views can be clearly seen in two early scientists, Descartes (1596-1650) and Robert Hooker (1635-1703) who lived at the height of the "Rosicrucian controversy" and who contributed much to science. They represent two sides of the onlooker, of the isolation of humanity as a "world spectator", two views as to the objective value of human thinking. Descartes brought to birth modern philosophy with his dictum "I think, therefore I am," by which he meant that the mind only contains images of sensory phenomena, and knows nothing of the objects themselves. Thus he felt the only thing in the universe he could count as real was his own thinking, since even to doubt the existence of it would require him to think. Hooker took the opposite position. Due to his work with the newly invented microscope, he discovered the cellular structure of plant tissues. But this work led him to conclude that the human being's sense-perceptions corresponded little to the world revealed by the microscope. This was the beginning of scientists' using instruments to give artificial sense impressions. It negated Descartes' idea that thinking was the one sure measure of existence, and in the subsequent philosophy of Hume, led to "universal skepticism."[33]

Science began to be mired between the unreality of concepts based on sense perceptions, and the equally unreliability of sense-perceptions themselves. The scene was set for Immanuel Kant and the strict limits he prescribed for human thinking which I described in an earlier chapter.

### Rosicrucian Influences

The publication of three documents in the first years of the seventeenth century caused a great furor; many people began to search feverishly for the Brothers of the Rosy Cross, whose organization announced its presence in Europe by these tracts[34] According to Steiner, the Rosicrucians had already been very active for some three centuries. He gave a description of these activities, which took three major forms: as healers, as medical men; as teachers of Mystery wisdom; and finally in a working behind the scenes to influence the social and political life.[35]

In healing, these men used techniques much like those of Paracelsus, a deep understanding of herbal medicaments and an exact knowledge of earthly-cosmic correspondences as they meet in the human being. For their services they took no money. As teachers of a profound Mystery wisdom, the Rosicrucians lived the simplest outward lives, often as herb-gatherers. Students had to search them out diligently, for they created no outward show of their great knowledge. These teachers taught through intense meditation on symbols, many of which have come down to modern times.[36] In their teaching methods, we can observe human consciousness in transition from the medieval mind to modern consciousness, with the strong reliance still placed on bringing the highest feelings to bear in the meditation on such symbols, often of geometrical design, yet with a determined study of "the book of nature." Thus art, science, and religion meet in the drawing and use of such symbols, and in the precise observation of outer nature.

To enter the social sphere, the Rosicrucians made use of the particular stage of human evolution. In medieval times, they inspired the *minnesingers*, the troubadours, who gave in picture-form great spiritual truths. Chief among these were the Holy
Grail legends, particularly the Parsifal cycle, which show in poetic images the soul striving for spiritual knowledge. Through such popular stories, they sought to inspire the whole culture to reach spiritual heights. The great cathedrals gave a lasting form in stone and glass to the medieval soul as it strived mightily upward to regain spiritual knowledge lost in the course of the Kali Yuga.

I have emphasized the key role of folk-wisdom in European history. Steiner showed how the Rosicrucian impulse led to the revitalization of this folk-wisdom through the perpetration of fairy tales, which gave important facts of spiritual wisdom in "mythopoeic" form.[37] Thus by their influence on the literature of the Middle Ages and its subsequent effect on the social, religious and artistic life, and their nourishment of the folk-tradition, the early Rosicrucians sought to aid the spiritual life of Europe.

With the dawning of the "spectator" consciousness, which Steiner called the "consciousness" or "spiritual" soul, these methods and tactics had to be changed. The great threat was that materialistic thinking would completely ensnare the coming scientific revolution. All the Rosicrucian efforts of the sixteenth and seventeenth centuries must be seen in light of this grave danger. The publication of the three documents, making the brotherhood a matter of public knowledge, represents a last-ditch effort to stem the tide of materialism and the mechanical world-view of the detached, passive onlooker-scientist.

The many societies of scientists and others that Yates describes so thoroughly as inspired by the Rosicrucians are, if Steiner is correct in his research, an attempt to let Mystery knowledge flow once again into European culture, especially into the natural sciences. It was a call for a third Reformation, not Protestant or Catholic, but based on brotherly love, a real knowledge of microcosmic-macrocosmic relationships, and on a "turning toward the works of God in nature in a scientific spirit of exploration" to perceive the working of the spirit there. Frances Yates wrote that "Rosicrucian thinkers were aware of the dangers of the new science, of its diabolical as well as its angelical possibilities, and they saw that its arrival should be accompanied by a general revival of the whole world."[38]

Before we turn to the possible influence of this Mystery school on agriculture, a brief word about the facts of the Rosicrucian efforts of the early seventeenth century, which fell under the horribly devastating Thirty Years War (1618-1648), for which many estimates hold that between the Italian Alps and the North Sea in Central Europe, nearly two-thirds of the chattel property, including livestock, and over one-half of the population perished. In addition to the war, vicious and recurrent "witch-scares" swept Europe. Frances Yates told of the philosopher Descartes's experience as a soldier during this war. One night, profound dreams convinced him that "mathematics were the sole key to the understanding of nature."[39] The witch-scare caused Descartes to deny any association with the Rosicrucians when he returned to Paris in 1623, after military service in Moravia, Siberia, northern Germany and the Catholic Netherlands, Yet he met one man, Johann Faulhaber, who wrote of the Rosicrucians. Yates relates Descartes's vigorous but, as he claimed, unsuccessful search for the Brotherhood. I can only agree with Yates that it is fascinating to speculate on what might have happened had he made a real connection. She surmised that it is quite possible that the man whose mathematics gave such a strong impetus to mechanistic and materialistic science yearned all his life for a more spiritual approach but failed to find it.

Yates concluded that "wars and witch-crazes have perhaps confused for the historian the vital steps by which the European mind moved out of the Renaissance into the seventeenth century."[40] Yet England escaped the devastation visited upon the Continent. Elizabeth defended religious freedom as did her successor. In Italy, even Galileo landed in prison for his view of Copernican astronomy. In England, people were free to form the many scientific societies that arose along Rosicrucian lines, chief among them the Royal Society. Isaac Newton, one of the greatest mathematicians, was a member of the Society and was very familiar with Rosicrucian writings. That he and Descartes obviously searched for more meaning than the spectator- consciousness could find in its studies of nature, and failed to find it, seems to me proof that a possible breakthrough of Mystery-wisdom was thwarted in the early seventeenth century. Not until the natural scientific work of Goethe over two hundred years later would another alternative method of inquiry into the natural world begin.

## The Agricultural Revolution

For many years historians generally agreed that agriculture experienced one of its periodic revolutions in the eighteenth century, resulting in an increased productivity which provided the food to support the later Industrial Revolution. Subsequent investigations are pushing the date of this period of agricultural innovation steadily into the past, and now estimate its inception somewhere in the late sixteenth and early seventeenth century. According to the new view, the great eighteenth century pioneers such as Coke, "Turnips" Townshend, Jethro Tull, and Robert Blakewell, refined, augmented and publicized methods which were already in use.[41] If the revisionists are correct in their assessment, then agricultural improvements began at the time of the ferment described above. After a brief discussion of the major trends of the revolution, I will turn to the question of the influence of the Rosicrucian impulse.

The main aspects of the Agricultural Revolution are: the enclosure of the common fields to make large pastures for livestock; replacing the bare fallow-field with root-crops such as turnips or with grasses; the four-crop rotation, with an increased use of legumes; the increased use of seed-drills and other improved implements; better drainage techniques; and improved the quality of livestock breeds. These changes came gradually, over a century and more. An important factor was the planting of turnips for fodder which in turn gave more manure for the grain crops. Mixed farming—the raising of both grain and livestock for market and dairy purposes—was pioneered in Holland in the sixteenth century and then brought to perfection in England.[42]

The Elizabethan poet, Sir Philip Sydney, was quite influential during his life and was a devoted Rosicrucian. He published the first English translation of the chief Rosicrucian manifesto, *The Chemical Wedding of Christian Rosenkreutz.* His pastoral poem, *Arcadia,* not only had utopian overtones as did many Rosicrucian writings, but concerned an intensive study of nature by the sensitive and innocent shepherd. In other words, we now must care for nature as a shepherd tends his flock. Pastoral literature, wrote J. Huizinga in *The Waning of the Middle Ages,* had been one of the strongest themes of that period. Paul Shepard said: "However artificial it might be, pastoral fancy still tended to bring the loving soul into touch with nature and its beauties. The pastoral genre was a school where a keener perception and a stronger affection toward nature were learned. . ."[43] It seems quite natural to me that Sydney should write in this genre, especially in light of the Rosicrucian activities in medieval literature discussed above. Literature flourished in the reigns of Elizabeth and James I, and it was a perfect medium to further a healthy view of nature, science, and religion.

Rosicrucian efforts to establish scientific societies are even more relevant than literary ones for agriculture. "Agricultural improvement" societies had a large influence on British farming by speeding the flow of scientific knowledge and new techniques throughout the land. Many members were also scientists. These societies quickly spread to America, where such leaders as Jefferson, Benjamin Franklin, George Washington and Patrick Henry showed a strong interest in scientific farming. Information spread rapidly through society publications and meetings, and by observations, as other farmers observed the innovators' methods. As Patrick Henry said, "Since the achievement of independence, he is the greatest patriot who stops the most gullies."[44]

The current state of knowledge makes it difficult to trace the many inventors and innovators, and the members of the various improvement societies, and I found no evidence that historians are aware yet of a possible Rosicrucian impulse behind seventeenth and eighteenth century agriculture.[45] Based on Steiner's description of the Rosicrucians, on Yates's scholarship, and on the pattern that has emerged in this study, I believe it likely, and an avenue for further research. The most fruitful areas for research would be, of course, England, and also the United States and the Netherlands.

Evidence exists of a strong Rosicrucian impulse in Pennsylvania among the German immigrants, and also some information linking Massachusetts' famous Adams family to English occultists.[46] it is a well known fact that very many early American leaders had a deep connection to Freemasonry, including Franklin, Jefferson, Dr. Benjamin Rush, Washington and Madison; and Yates discussed the influence of Rosicrucian thought on the origins of the Masonic order. She also found that the British Puritan John Winthrop, a follower of Dr. John Dee, brought Rosicrucian alchemy to New England.[47] Given the state of current knowledge and the limited space available for discussion here, I can do no more than point out these intriguing connections and questions for others to explore. Today historians delve ever deeper into local records in political, social, and agricultural history. Hopefully evidence will come to light that documents the influence of the Rosicrucian impulse in the formation of agricultural societies in Great Britain and the United States. Yates and Steiner have established the pattern and motive. How historians must do the leg-work.

## Agriculture in the Late Eighteenth and Nineteenth Centuries

Throughout the eighteenth century, progress continued in agriculture. The major advances came in England for many reasons. The gentleman scientist-farmer took the entrepreneurial spirit from commerce and industry into farming. England had far fewer class-tradition restrictions than did the continent, great though they remain even today, and was spared the devastation of war on her lands. The same influences that began the agricultural revolution also fired the well known Industrial Revolution, which in turn benefited farming. Several industrial wastes, most notably lime, made excellent fertilizer. Eric Kerridge wrote: "Every advance in industry, indeed, swelled supplies of extraneous fertilizers." The increasing use of coal allowed more lime for the fields, and more manure and straw for compost, since they were not needed for fuel. Soap-ash residue from lye-making helped clover to grow.[48]

Some statistics will illustrate the progress of agriculture. The average yields increased dramatically. B. H. Slicher van Bath found that between 1200-1249, farmers harvested 3.7 gains for every one sown; between 1500-1700, 7 grains for every one sown; between 1750-1820, 10.6 grains harvested for one sown. In England, between 1701-1801, the population grew by 6.5 million, or over 70%, supported by the agricultural revolution. Late changes were even more dramatic. French farmers, called the poorest in Western Europe, increased their productivity by 50% between 1850 and 1880.[49]

Most of these innovations had salutary effects on agriculture. As the nineteenth century began, the first signs of trouble appeared, although progress covered them in her mantle of success; but for the first time, the criteria of industrial production began to be applied to agriculture. A rather rapid change ensued: the farmer more and more became nature's manipulator instead of the mediator between her forces. With increasingly better machinery, the riches of nature could be exploited even more ruthlessly. England, and later Europe, would have ruined her land in a short time since there was only a limited amount, but she found it possible to buy grain from her former colonies in America. While this hurt English agriculture financially, it saved the soil from much greater damage.

The scientific world-view by 1925 had emerged from the hermetical-Rosicrucian doctrines almost completely; animist and vitalist views fell before the advance of mechanistic and materialistic theories. No serious scientist spoke of the "soul" of the world, and few even recognized any concept of a life force, of an organizing principle above mechanical and inorganic chemical laws. The noted chemist, Baron Justus von Liebig began to apply inorganic chemistry to biology and founded "organic" chemistry. It is most important to note that organic chemistry treats the processes of living things no differently than inorganic chemistry does the combinations of minerals. It is "organic" in that the compounds it isolates from living things are usually much more complicated in structure than mineral molecules; but the organic chemist's mind does not attain to the level of *life* in his work; thus, he divides the plant or animal into its chemical constituents. The great danger is that he all too easily regards the living, whole being, which is in turn inextricably wedded to the whole of life, as only the sum of its chemicals, of its parts.

Baron von Liebig, one of the great chemists, made the fundamental discovery that plants feed on mineral salts in solution. Photosynthesis was not known, nor was the function of trace elements, but Liebig and others found the importance of nitrogen, phosphorus and potassium, present in the soil as nitrates, phosphates, and potash, now universally called "NPK". For thousands of years men had wished they could "fatten" field crops as they did their livestock. Von Liebig gave them the means— chemical fertilizers to replace the mineral elements the plants took from the soil. But now the gate was open to grow crops in total disregard for the true "life" of the soil, the humus and the billions of micro-organisms which made humus and transformed it into compounds the plants can use. No longer did farmers have to depend on animal production of manures, complicated crop rotations composting to add organic matter to the soil. Now the great power and skill of industry could manufacture seemingly endless supplies of chemicals. Farmers could kill their soil without the warning of crop failures. As long as chemical fertilizer was available, only such disasters as the Oklahoma Dust Bowl revealed the lifeless soil, which merely supported plants that grew in it by a sort of "hydroponics."[50]

Here we see the onlooker consciousness at its best, or worst, depending on one's perspective. The farmer could emancipate himself quite considerably from the cycles of nature. He could "mine" his soils of their minerals, sell the produce to industrial cities, buy industrial chemicals for his land, and continue on in this manner. Millions of acres in Europe escaped this destruction, because it was cheaper to buy produce from the United States and South America, whose soils, with chemical help, continue to produce prodigious quantities of food. Only today have people begun to question the damage to soil and nutrition.

Interestingly, Baron Justus von Liebig, whose *Chemistry Applied to Agriculture and Physiology* and *Natural Laws of Husbandry* went through many editions and translations, knew that the inorganic mineral nourishment of plants was only part of the story. He was no narrow specialist, nor was he totally materialistic in his thinking, for he wrote:

> Inorganic forces breed only inorganic substances. through a higher force at work in living bodies, of which inorganic forces are merely the servants, substances come into being which are endowed with vital qualities and totally different from the crystal. . . The cosmic conditions necessary for the existence of plants are warmth and light of the sun.[51]

Agricultural education during the late nineteenth century fell under the increasing domination of the university. The old folk-wisdom, while it remained strong in sections of Europe, gave way to heavily mechanized chemical farming in the United States. On both continents agricultural scientists began to set the tone for the farmers' thinking. Agricultural colleges were established in America in 1862, and in Europe even earlier. The "professor", the teacher concerned only with the head and its thoughts, thus rose to the fore in agricultural education. No longer concerned with humanity's connection to the whole world, as had been the "gymnast" and the "rhetorician"; the professor symbolizes the teacher cut off from any idea of microcosmic-macrocosmic relations—the passive, detached onlooker.

**New Directions**

In the early years of this century, several people began to look into farming that did not destroy the soil structure or the life within it, two of the most effective being Sir Albert Howard and J. I. Rodale.[52] These men sought a thoroughly scientific approach to organic agriculture, and they developed many useful things, including excellent composting techniques. Howard himself was both a fine farmer and an enterprising experimenter, and his methods are used on farms in Ceylon, India, South Africa, Central America, and the United States.

Yet what of the last sentence of von Liebig, quoted above? What of the cosmic forces? Knowledge of humanity's place in the cosmos, of celestial influences on humanity's soul and on the earth—this had always been preserved, nurtured and taught in the Mysteries. And the Mystery wisdom had long since been regarded as superstition by the materialistic world-view. Even those people who sought consciously to find some meaning in the old symbols and philosophies found little to sustain them. As early as the late eighteenth century, the old streams were rapidly fading. Writing later in 1847, Rothe of Heidelberg, a student of esotericism wrote:

> At all times there have been very few in whom this insistent speculative need has been combined with a living religious need. But theosophy is for these few alone . . . The important thing is that if theosophy ever becomes scientific in the real sense and produces obvious and definite results, these will gradually become the general conviction, be acknowledged as valid truths and be universally accepted by those who cannot find their way alone, the only possible path by which they could be discovered. But all this lies in the womb of the future which we do not wish to anticipate.[53]

Rothe saw that theosophy, by which he meant the old Mystery wisdom, would have to enter the field of science once again, which brings to mind Cicero's statement that "Those occult Mysteries . . . when interpreted and explained have more to do with natural science than with theology."[54]

In a previous chapter I discussed how Goethe forged a new beginning in scientific inquiry, which Rudolf Steiner gave a thorough epistemological foundation. Goethe himself felt a great affinity to the old wisdom, and in his poem The Mysteries and in the story of *The Green Snake and the Beautiful Lily*, he attempted to recast certain truths in a form suitable for the modern mind.[55] He had been deeply influenced by Rosicrucian circles as a young man.

Rudolf Steiner gained much help in his insights into nature from his meetings with the old Austrian herb-gatherer I mentioned earlier. This man also led Steiner to another person who through the philosophy of Fichte had developed observations which profoundly affected Steiner's later work. In these two men we can see the submerged stream of Mystery wisdom rise to the surface of history to touch the life of Rudolf Steiner, just as it had done earlier in the case of Goethe. The results for the development of agriculture were far reaching, as was the case for other aspects of cultural life, most notably education, philosophy, and medicine.

**Summary**

In this section, I have presented Steiner's concept of the development of human consciousness as a background for the history of agriculture. In the first section of the study, I described the epistemology of Steiner, in which thinking becomes the activity by which humanity can gain knowledge of the spiritual world as it manifests in nature, in the sensory-physical world. In this second section, we have seen how the development of such enhanced cognition has always been the concern of mystery centers and their initiates, and I presented Steiner's ideas which indicate the close relation between these centers and agriculture. I gave a short description of Steiner's concept of the central role of the being of Christ in human evolution, and how those concerned with Christian esotericism turned to the world and sought to redeem both humanity's thinking and the earth itself. This development reached a certain culmination with the Rosicrucian activities of the seventeenth century. Since that time, the Mysteries have been eclipsed by the scientific world-view.

Perhaps the chief Mystery teaching was of the connection that exists between the earth and the cosmos, between spirit and matter, between the human soul and the world. Since the Mystery of Golgotha, the development of "I-ness" or ego consciousness has become responsible for receiving the Higher Self, the Christ, into the lower self. This, we saw, took place on the psychological level in the exclusive Mystery rites, but now can take place in every human soul. Yet as the ego-consciousness achieved its highest degree of individuality, it also lost sight almost completely of the old wisdom, of the earthly-cosmic correspondences. Spiritual knowledge became tradition and dogma to be accepted on faith.

Rudolf Steiner held that the great five thousand year cosmic rhythm, called the Kali Yuga, ended in 1899, and a new cycle began. In the new age, now nearly a century old, humanity will gradually regain its clairvoyant powers. The new cognition, he taught, would build on the powerful thinking and inner logic developed as a faculty along with ego-consciousness. Thus he called his own philosophy "spiritual science." Steiner's work in a certain sense, then, is a re-formulation of the ancient mystery wisdom into a body of knowledge thoroughly in accordance with the natural sciences, but extending beyond the boundaries of knowledge recognized by modern science.

Steiner's concept of enhanced cognition is a modern path of knowledge which brings the old Mystery wisdom fully into the light of ego-consciousness. Yet the "birth" in the Mystery of the Logos in the soul still must take place, whether through faith or through self development. Once again, he said, humanity must

use his soul powers actively, not passively as a spectator. Once again humanity must become aware of earthly-cosmic correspondences and must let such knowledge guide his affairs. Steiner
was convinced that human freedom and health depended on it. Against this grand sweep of human evolution, Steiner's indications for a new agriculture were born. As he said, speaking of the necessity for a new agriculture and science to repair the damage done by current practices:

> . . . a decrease of the products is observable. The decrease is indeed connected—like the transformation in the human soul itself—with the ending of the Kali Yuga in the Universe during the last decades and in the decades that are now about to come. You may take my remark amiss or not, as you will. We stand face to face with a great change, even in the being of Nature. Whatever has come down to us from ancient times—whatever it may be that we have handed down: natural talents, knowledge derived from Nature, and the like, even the traditional medicaments we still possess—all this is losing its value.
>
> We must gain new knowledge in order to enter again into the whole nature-relationship of these things. Mankind has no other choice. Either we must learn once more, in all domains of life—learn from the whole nexus of Nature and the Universe —or else we must see Nature and withal the life of Man himself degenerate and die. As in ancient times it was necessary for men to have knowledge entering into the inwardness of Nature, so do we now stand in need of such knowledge once again.[56]

**Notes for Chapter Five**

1. Rudolf Steiner, *An Outline of Occult Science* (New York: 1972) 250.
2. Ibid., 253-254.
3. Ibid., 252. See also Eleanor C. Merry, *The Flaming Door: A Preliminary Study of the Celtic Folk-soul by Means of Legends and Myths,* rev. ed.(East Grinstead: New Knowledge Books, 1962); Jakob Streit, *Sun and Cross: The Development from megalithic culture to early Christianity in Ireland* (Edinburgh: Floris Books, 1984); Christopher Bamford, ed. *Celtic Christianity* ( W. Stockbridge, MA: Lindisfarne Press, 1982); and Ean Begg, *The Cult of the Black VIrgin* (London: Arkana, 1986).
4. This discussion comes from the important study by Lynn White Jr. *Medieval Technology and Social Change* (London: Oxford University Press. 1962) 39-79.
5. For the medieval cattle situation see G.E.Fussell, *Farming Techniques !from Prehistoric to Modern Times* (Oxford: Pergamon Press, 1965) 96-101.
6. Charles Parain, quoted in Lynn White Jr., *Medieval Technology and Social Change*, 69.
7. White, Ibid., 78.
8. Lynn White Jr. "The Historical Hoots of Our Ecological Crisis, *Science*, 155(10 March 1967) 1205.
9. Rudolf Steiner, *The Tension Between East and West* (London: Hodder and Stoughton, 1963) 52.
10. See Steiner's discussion in *World History in the Light of Anthroposophy* (London: 1950) 103-107.
11. Steiner, *World History,* 104. A more lengthy discussion will be found in Steiner, *Human Values in Education* (London:1971) in the last chapter.
12. Steiner, *World History*, 105-106; Steiner, *The Redemption of Thinking: The Philosophy of Thomas Aquinas* (London: Hodder and Stoughton, 1956); and Steiner *Mystery Knowledge and Mystery Centres*, 2nd ed. (London: 1973) chapter 13.

13. Rudolf Steiner, *Rosicrucianism and Modern Initiation*, 2nd. ed. rev. (London:1965) 15; Lynn White Jr. "The Historical Roots of Our Ecological Crisis", 1206.
14. Some of my discussion of the Cistercians and Chartres comes from a series of lectures by René Querido given at the Waldorf Institute, Detroit, Michigan, 24-25 November 1977. Querido is a long-time student of Steiner's work. The comment on Christian esotericism comes from Rudolf Steiner, "European Mysteries and their Initiates," *Anthroposophical Quarterly* (Spring, 1964) 173. See René Querido, *The Golden Age of Chartres: The Teachings of the Mystery School and the Eternal Feminine* (New York: 1989).
15. See Louis Charpentier, *The Mysteries of Chartres Cathedral* (New York: Avon Books, 1966); Rudolf Steiner, *The Evolution of Human Consciousness* (London: 1966); and Steiner, *Mystery Knowledge and Mystery Centres*, lectures 7, 8, and 9. See also René Querido, *The Golden Age of Chartres: The Teachings of the Mystery School and the Eternal Feminine* (New York: 1989) and his "The Cathedral and the Great Masters of Chartres," *Journal for Anthroposophy*, (Issues 45 and 46), and Joel Morrow, "A Thread from the Tapestry Alanus Wove: Nature and Inner Development in Alan of Lille and Bernardus Silvestris," *Journal for Anthroposophy*, 51(Fall 1990).
16. See notes 14 and 15.
17. Louis Gottschalk, Loren C. McKinney, and Larl C. Pritchard, *The Foundations of the Modern World* (London: George Allen & Unwin, 1969) 952; Charles Parain, "The Evolution of Agricultural Techniques in the Middle Ages," *Cambridge Economic History of Europe* I (London: Cambridge University Press, 1966) 135, 129; and Richard Kroebner, "The Settlement and Colonization of Europe," Ibid., 75-76, 298, 359.
18. The quote is from Henri Pirenne, *The Economic and Social History of Medieval Europe* (NewYork: Harvest Books, 1937) 69. See also pages 68, 75, 77, 151. Parain, Ibid., 130.
19. From a conversation with René Querido, Waldorf Institute, Detroit, Michigan, 25 November 1977.
20. Rudolf Steiner, *World History*, 103-104.
21. G.E. Fussell, *The Classical Tradition in Western European Farming* (Rutherford: Farleigh Dickinson University,Press, 1972) 79.
22. Stewart C. Easton, *Man and World in the Light of Anthroposophy* (New York: 1975) 95. Hereafter cited as *Man and World.*
23. Ibid., 57-58.
24. Ernst Lehrs, *Man or Matter: An Introduction to a Spiritual Understanding of Nature on the Basis of Goethe's Method of Training Observation and Thought* , 2nd ed. rev. (London: Faber and Faber, 1958) 47-48.
25. Owen Barfield, *Saving the Appearances: A Study in Idolatry* (New York: Harcourt Brace & World, n.d.).
26. Frances A. Yates, *Giordano Bruno and the Hermetic Tradition* (New York: Vintage Books, 1964) hereafter cited as *Giordano Bruno*; and Yates, *The Rosicrucian Enlightenment* (London: Routledge and Kegan Paul, 1972).
27. Rudolf Steiner, *Mysticism at the Dawn of the Modern Age* (New York:1982)..
28. Yates, *Giordano Bruno*, 407-408.
29. Yates, *The Rosicrucian Enlightenment*, 219.
30. Paul M. Allen, ed. *A Christian Rosenkreutz Anthology* (Blauvelt: Rudolf Steiner Publications, 1968) 456. See also Peter French, *John Dee: The world of an Elizabethan magus* (London: Routledge and Kegan Paul, 1972) and Frances A. Yates, *The Occult Philosophy in the Elizabethan Age* (London: Routledge and Kegan Paul, 1979), particularly parts 2 and 3.
31. Rudolf Steiner, *The Mission of Christian Rosenkreutz* (London: 1950); and many other lecture cycles. For example, *Rosicrucianism and Modern Initiation* , and lectures 13 and 14 of *Mystery Knowledge and Mystery Centres.*
32. Steiner, *Occult Science*, 248; see also Ren´Querido and Hilmar Moore, *Behold, I Make All Things New: Toward a World Pentecost* (Fair Oaks: Rudolf Steiner College Publications, 1991).
33. Ernst Lehrs, *Man or Matter*, 50-51.
34. See Yates, *The Rosicrucian Enlightenment* and Allen, *A Christian Rosenkreutz Anthology.*
35. The three documents, or manifestoes, are: The Chemical Wedding of Christian Rosenkreutz; Fama Fraternitatis; and Confessio Fraternitatis. They can be found in Paul M. Allen, Ibid..

36. For many of the Rosicrucian symbols see Ibid.
37. Rudolf Steiner, *Background to the Gospel of St. Mark* (London: 1968) 178-179.
38. Yates, *The Rosicrucian Enlightenment*, 117.
39. The phrase is Frances Yates'. Ibid., 139. 233.
40. Ibid., 119.
41. See for example G.E.Mingay, "The Agricultural Revolution in English History: A Reappraisal", *Agricultural History* 26(1963) 123-133; Eric Kerridge, *The Agricultural Revolution* (London: George Allen & Unwin, 1967); and E.L. Jones, *Agriculture and the Industrial Revolution* (New York: Halstead Press, 1974) 67-70.
42. Kerridge, Ibid.,15; D.B.Grigg, *The Agricultural Systems of the World: An Evolutionary Approach* (London: Cambridge University Press, 1974) 165-167.
43. Paul Shepard, *Man in the Landscape: An Historic View of the Esthetics of Nature* (New York: Knopf, 1967) 75-76.
44. Patrick Henry quoted in Vernon Gill Carter and Tom Dale, *Topsoil and Civilization*, rev. ed. (Norman: Oklahoma University Press, 1974) 228; F.L. Jones, *Agriculture and the Industrial Revolution*, 118-119; and David Brandenburg, "Commentary on Eighteenth Century British Agriculture," *Agricultural History* 42(1974) 19-24.
45. See for example Eric Kerridge, *The Farmers of Old England* (Tobowa, N.J.: Rowman and Littlefield, 1973) 130.
46. For Pennsylvania see Allen, ed. *A Christian Rosenkreutz Anthology*, 215-217, For the Adams family see, see Andrew E. Hothvius, "The Dragon Tradition in America," *East West Journal* (May, 1977) and (August, 1977).
47. Yates, *The Rosicrucian Enlightenment*, 206-219; 227 n.l.
48. Kerridge, *The Agricultural Revolution*, 242-243, 247-248; E.L. Jones, *Agriculture and the Industrial Revolution*, 90, 93.
49. Slicher van Bath quoted in Fernand Braudel, *Capitalism and Material Life, 1400-1800* (Hew York: Harper & How, 1973) 275; J.D. Charters and G.E. Mingay, *The Agricultural Revolution, 1752-1880* (London: B.T. Batsford, 1961) 173.
50. Edward Hyams, *Soil and Civilization* (New York: Harper Colophon Books, 1976) 270-271.
51. Quoted in E.E. Pfeiffer, "Preface", to Rudolf Steiner, *Agriculture,* 3rd. ed. (London: Biodynamic Agricultural Association: 1974) 9.
52. Among several books by Howard, see *The Soil and Health: A Study of Organic Agriculture* (New York: Schocken Books, 1972). See Hilmar Moore, "Liberty Hyde Bailey and the Holy Earth," *Journal for Anthroposophy,* 40/41 (Spring 1985).
53. Steiner, *Background to the Gospel of St. Mark*, 168-169.
54. See my chapter four, above, note 1.
55. J.W. von Goethe, "The Mysteries," *Journal for Anthroposophy*, 25(Spring, 1977) 32-40; and his *The Green Snake and the Beautiful Lily* (New York: Knopf, 1971).
56. Rudolf Steiner, *Agriculture*, 39.

## PART THREE:

## RUDOLF STEINER'S APPROACH TO AGRICULTURE AND EDUCATION

# CHAPTER SIX

# BIODYNAMIC AGRICULTURE

*The environmental crisis is a sign that the finely sculptured fit between life and its surroundings has begun to corrode. As the links between one living thing and another begin to break down, the dynamic interactions that sustain the whole have begun to falter and, in some places, stop.*
*Barry Commoner*[1]

### The "Agricultural Course"

From June 7 to 16, 1924, some sixty farmers, veterinarians, scientists, and gardeners attended a series of lectures and discussions with Rudolf Steiner, subsequently known as "the agricultural course", and today published as *Agriculture*.[2] This meeting, which took place on the large farming estate at Koberwitz of Count Carl von Keyserlingk, followed previous researches and practical applications conducted under Steiner's supervision at the Bio-Chemical Research Laboratory in Dornach by Guenther Wachsmuth and E. E. Pfeiffer in 1921-1923. The interest generated by these scientific efforts led Keyserlingk to offer his estate for an agricultural meeting, but Steiner's constant activity—tours, lectures, writing, meetings—caused him to postpone his consent to the course several times. finally, the Count sent his nephew to Dornach with the instruction to sit on Steiner's doorstep until he received Steiner's firm commitment; thus the dates were settled.[3]

During the last year of his public activity, from 1 September 1924 to 23 September 1924, Steiner worked on his autobiography, gave definitive lecture courses on speech eurythmy, tone eurythmy, speech and drama, medicine, theology and severs, and education. Among the cities he visited were Prague, Berlin, Bern, Arnhem, Torquay, Stuttgart, London, the Hague, and Vienna. He also gave major cycles on world history, Mystery centers, evolution, and "pastoral medicine" for priests and doctor. In the last three weeks of this period, he gave seventy lectures, often three and four per day, in addition to his writing, preparation, administrative duties, and a nearly endless stream or personal interviews. For long periods he slept only a few hours each night.[4]

The agricultural lectures thus came in the midst of intense activity, and the request for them arose from several directions, including the veterinary Dr. J. Werr, Dr. Eugen Kolisko, a physician, and staff members of the Weleda Company, a pharmaceutical firm established to produce medicines and cosmetics according to Steiner's indications. They asked for help in treating the increasingly prevalent animal diseases. Farmers such as Ernst Stegemann inquired about the varietal degeneration of seed stocks and many cultivated plants, and of proper farm management. The scientists Pfeiffer and Wachsmuth sought to research the formative life-forces of plants. In his lectures, Steiner gathered his previous indications to these people and presented a concentrated course as "an introduction to understanding spiritual, cosmic forces and making them effective again in the plant world."[5]

Steiner sought to give agriculturalists a "fundamental re-orientation" with which to meet the ever-growing problem caused by the modern scientific world-view, and the increasing domination of farming by the money-market, with its essentially industrial,rather than agricultural, outlook. Compared to the massive problems that agriculture faces today, the questions asked of Steiner may seem of perhaps minor importance; mixed farming, crop rotations, and only the first stages of mechanization characterized the majority of farms in the early twenties. Yet these very problems have increased drastically and are at the root of today's troubles. E. E. Pfeiffer wrote of Steiner's agricultural work that:

> He never proceeded from preconceived abstract dogma, but always dealt with the concrete given facts of the situation. There was such germinal potency in his indications that a few sentences or a short paragraph often sufficed to create a farmer's or a scientist's whole life-work; the agricultural course is full of such instances. A study of his indications can therefore scarcely be thorough enough. One does not have to try to puzzle them out, but can simply follow them to the letter.[6]

## A Living Cosmos, A Living Earth

Steiner spoke of the earth as a living organism, part of a living cosmos. Within the cosmos exist two great polarities which pervade all of life's varied phenomena. These polarities are the cosmic forces, which stream in toward the earth from the periphery of the universe, and the earthly forces, which radiate outward from the earth's center. The centrifugal earthly forces, such as gravity, electricity, and magnetism, act between material centers, The cosmic forces, by contrast, are not centric forces, but are planar in nature a plane can be seen as an infinitely expanded point or sphere. Steiner called these peripheral forces the formative forces or life forces, or the etheric forces, They are the opposite of centric gravity; rather they are the forces of buoyancy and levity.[7]

In his essay "What is a Farm?" C.A. Meir wrote

> We shall never understand life by means merely of the centrifugal earth-forces. We must extend our view into this realm of the Cosmic forces, and then many a phenomenon which stands before us as a puzzle, or as an isolated fact or problem, will harmoniously fall into place as part of the tapestry of nature which, in all its varied manifestations, confronts the farmer day by day. Astronomy and geology, chemistry and botany, take on new light and color, and become much more accessible to the farmer.[8]

Specifically, Meir meant that the life principle present within all plants and animals—the forces that build the plant out of the mineral substances—this principle should be used to balance the one-sided centric view of the natural sciences.

The biodynamic philosophy views the plant between a polarity of forces. In addition to the earthly forces that provide substance, the cosmic forces stream in and supply the living processes of growth, nutrition, propagation, and the characteristic form-building of the plant. The farmer or gardener attempts to observe and work with these active processes, to see the plant dynamically, as it grows between "gravity" and "levity", as it manifests Goethe's law of metamorphosis in alternating stages of expansion and contraction. Scientists and farmers working out of Steiner's indications have brought these "Goethean" techniques of study to a high degree of sophistication.[9]

I discussed in previous chapters the gradual loss of the knowledge of such earthly-cosmic connections, culminating when men looked through the telescope and spectroscope and discovered in the stars the same elements and substances as exist on the Earth.[10] Pointing to the necessity of establishing "a genuine science, a science that looks to the great cosmic relationships," Steiner said:

> Nature and the working of the Spirit throughout Nature must be recognized on a large scale, in an all-embracing sphere. Materialistic science has tended more and more to the investigation of minute, restricted spheres. True, this is not quite so bad in Agriculture; here they do not always go on at once to the very minute—the microscopically small, with which they are wont to deal in other sciences. Nevertheless, here too they deal with narrow spheres of activity, or rather, with conclusions which they feel able to draw from the investigation of narrow and restricted spheres. But the world in which man and the other earthly creatures must live cannot possibly be judged from such restricted aspects.
>
> To deal with the realities of Agriculture as the customary science of today would do, is as though one would try to recognize the full being of Man from the little finger or from the lobe of the ear and trying to construct from thence the total human being.[11]

### Goethean Science and Biodynamics

A first step toward complementing the increasingly detailed search for causes of biological and chemical activity into microscopically small particles that characterizes modern research, is the Goetheanistic method of observation. In this method, the researcher attempts to create a real relationship with living phenomena in order to observe the formative processes at work, As H. H. Koepf wrote

> A study of the world of form revealed to the eye by living nature can lead us further. In the plant kingdom, and in another way also in the animal kingdom, we find phenomena in which certain formative principles become manifest in the material world in untold variety. The formative principle of all roses is the same, and yet each one is different. The relationship of the parts to the whole form also is not meaningless, because they belong together in their totality. We are not here seeking to understand the genotype as the alleged cause of the form—this lies on another level of consideration—but rather the form as an independent, spiritual reality. A study of the forms and their metamorphoses through exercises leads gradually to a spiritual participation in the creative, forma-

tive principle. On this path of study through exercises we learn step by step to read the gestures of nature by becoming spiritual sharers in them.[12]

The consequences of the modern scientific world-view, in which the researcher is in constant danger of losing sight of the object itself as it is in nature, become increasingly apparent as the fruits of such research are applied in practice. The biologist Barry Commoner maintains that "this same fault lies behind every ecological failure of modern technology: the attention to a single fact of what in nature is a complete whole."[13] The biodynamic farmer attempts, through Goethean observation and thinking to consider the plant itself in its relation to the entire plant community with its micro-organisms, insects, and animals. He looks beyond even this complex ecosystem into the heavens. His study of the creation of form also affords insight into the processes of decomposition of living substances; he learns to discern "in the growing and dying of plants, typical gestures of substance transformation."[14]

Such Goethean methods of approach provide a means to understand Steiner's lectures on agriculture and his other natural scientific work. Steiner's own powers of observation and intuition went far beyond Goethean science, and his agricultural ideas come directly out of his intimate knowledge of nature's secrets, which was a product of his highly trained and powerful clairvoyance. Only one with similar perceptive abilities could see how he arrived at his indications, which I will outline in this chapter. However, a healthy understanding and judgment, augmented by the Goethean approach, allow the farmer or scientist to apply Steiner's ideas and procedures with success and a deepening capacity for creative thinking.15 We can recall here the words of the agricultural scientist E. E. Pfeiffer:

> . . . the Biodynamic Method, however, can be employed by every farmer. This must be stressed, since many, later, got the idea that one could not work biodynamically if one were not an anthroposophist. That on the other hand knowledge of the biodynamic method gradually leads the one who uses it to another world-picture, that he to begin with learns particularly to judge the biological (i.e. life) processes and interconnections in a different way than the materialistic, chemically inclined farmer, is of course something obvious. So too, he will bring to the dynamic (i.e. play of forces) in nature a greater degree of interest and of conscious awareness. One must learn to understand that there is a difference between mere application of the method and creative collaboration.[16]

## Biodynamics and Cultural Renewal

*Before the late 1700s there was probably no settled community in which at least nine-tenths of the population were not directly engaged at tillage. Rulers and priests, craftsmen and merchants, scholars and artists, were a tiny minority of mankind standing on the shoulders of the peasants.*

*Lynn White, Jr.*[17]

Rudolf Steiner gave his agricultural lectures out of his spiritual vision into

the workings of nature. He hoped that they would enkindle a new relationship between humanity and the world of nature, a relationship that would be based on a perspective that goes beyond even the ecological sciences of today, which encompass the global ecosystems, out into the vastness of the cosmos. Steiner did not want to give merely a theoretical philosophy to be considered and debated intellectually; rather he sought
to present ideas and concepts only in order that they may become as vital within us, on the spiritual plane, as our life's blood itself, so that inner activity, not only thinking, is stimulated. A philosophy of life in accord with spirit thus reveals itself as a social as well as a cognitive impulse.[18] I believe it helpful to see Steiner's agricultural work in the context of his attempt to provide a manifold basis for cultural renewal.

Steiner's indications represent a foundation out of which new knowledge of nature can proceed, in which the old peasant intuition, fostered by the Aristotelian folk-tradition, can he reborn in modern, scientific agriculture. Today's social problems, he said "are symptomatic of the loss of the old instinctive certainties of social life and of the necessity to establish, consciously, a spiritual life that will give the same impulses as the earlier instinctive one."[19] Steiner voiced this concern for a new approach to life processes, for a more harmonious relationship with the living world, in his agricultural lectures:

> I grew up entirely out of the peasant folk, and in my spirit I have always remained there. I indicated this in my autobiography. . . I myself planted potatoes, and though I did not breed horses, at any rate I helped to breed pigs. And in the farmyard of our immediate neighborhood I lent a hand with the cattle. These things were absolutely near my life for a long time; I took part in them most actively.
>
> Therefore I beg you to consider me as the small peasant farmer who has conceived a real love for farming; one who remembers his small peasant farm and who thereby, perhaps, can understand what lives in the peasantry, in the farmers and yeoman of our agricultural life.
>
> For I have always had the opinion (this was not meant ironically, though it seems to have been misunderstood) I have always had the opinion that their alleged stupidity or foolishness is "wisdom before God", that is to say, before the Spirit. I have always considered what the peasants and farmers thought about their things far wiser than what the scientists were thinking.
>
> I have always been glad when I could listen to such things, for I have always found them extremely wise, while, as to science—in its practical effects and conduct I have found it very stupid. This is what we at Dornach are striving for, and this will make our science wise—will make it wise precisely through the so-called "peasant stupidity. We shall take pains at Dornach to carry a little of this peasant stupidity into our science. Then this stupidity will become— "Wisdom before God."
>
> Let us then work together in this way; it will be genuinely conservative, yet at the same time a most radical and progressive beginning. And it will always be a beautiful memory for me if this Course becomes the starting point for carry!ng some of the real and genuine peasant wit into

the methods of science.[20]

Rudolf Steiner said that he had wanted to gather the wisdom of the old folk-traditions into a coherent body of knowledge; perhaps, we may surmise, in a similar manner to his earlier philosophical work on Goethe:

> When I was a young man I had the idea to write a kind of "peasants' philosophy," setting down the conceptual life of the peasants in all the things that touch their lives. It might have been very beautiful.

He went on to state:

> A subtle wisdom would have emerged—a philosophy dilating upon the intimacies of Nature's life—a philosophy contained in the very formation of the words. One marvels to see how much the peasant knows of what is going on in Nature.
>
> Today, however, it would no longer be possible to write a peasants' philosophy. These things have been almost entirely lost. It is no longer as it was forty or fifty years ago. Yet it was wonderfully significant; you could learn far more from peasants than from the University.[21]

Steiner thus envisioned a rigorously practical agricultural approach, one which requires both the intuition of the vanished peasant farmer and the inner logic and intellectual understanding of the scientist, and which gives the opportunity to collaborate actively and creatively with the life processes of nature. The biodynamic farmer can learn to achieve a "clear and conscious relationship to the world of forces appearing in living organisms, thereby giving new meaning and dignity to his work, and alleviating the ennui and alienation so prevalent today."[22] Steiner's indications afford the farmer and gardener practical methods to foster common sense, intuitive insight and instinctive certainty, and to mediate between earthly and cosmic processes and rhythms. To be applied with optimum effect, the methods demand that one's agricultural development and one's self-development must find a fruitful marriage. The growth of plants, the maintenance and increase of soil fertility become tied to the farmer's inner growth.[23] As the farmer enhances his perceptions and enlivens his thinking, he begins to perceive the forces of life pulsating through the plants and animals. The biodynamic method allows him to increase the life in his fields and forests.

## Some Biodynamic Principles

The biodynamic method seeks to diversify farms in accordance with the demands of the local environment. Crops are balanced between those that build fertility and those that exhaust the soil. Nitrogen-fixing legumes are planted extensively, and the whole emphasis is on production based on stable fertility. Biodynamic farms are organized as "balanced biological units;" in other words, a proper number of livestock are raised to furnish most of the manure for composting. Manures and other organic wastes are composted to preserve and enhance their fertilizing capabilities for increased soil life. *The biodynamic preparations* improve the life-building qualities of the compost, increase soil life by direct spraying, and stimulate the plants' utilization of light. Six of the preparations are made from plants which are enclosed in specific animal organs and exposed during certain seasons in soil or atmosphere to the prevailing, environmental influences. The plants used are valerian, yarrow, chamomile, dandelion, oak bark, and nettle; two other preparations are made from quartz and cow dung, respectively. Steiner gave the substances and animal organs for the "specific contribution each could make to the composting process" or to soil and plant life.[24]

In addition to these measures, biodynamic farmers and particularly gardeners make use of *companion planting*, another example of discerning and utilizing the myriad interrelationships in nature. The study of plant symbiosis Is a relatively "new" field, although the folk traditions have preserved for centuries examples of plants that grow well or poorly together. In addition to the better known interaction by competition for water or nutrients or light, many plants actually help others to grow. Some plants, such as alfalfa or dandelion, require soil with deep root systems, providing paths through which earthworms may tunnel, breaking the hard pan and enormously increasing the relation between subsoil and topsoil, and allowing the cosmic forces of light and warmth to penetrate more deeply. Other plants enrich the soil with minerals and organic substances, drawn up through the roots or "fixed" from trace elements in the air. Steiner spoke of the importance of trace elements, enzymes, and other biocatalysts as early as 1924; since then, a growing body of research has confirmed his views.[25] Some plants excrete odors which repel or attract insects, or other helpful compounds. Often the relationships are subtle, and experiments have trouble in explaining why they work. However, biodynamic researchers and practitioners have assembled a large quantity of information on this subject.

Perhaps a few examples will illustrate companion planting: stinging nettle (Urtica dioica) when planted near certain herbs, such as lavender, rosemary, and most especially peppermint, stimulates their pungency and aroma by increasing the content of their essential oils. Experiments have shown increases of these oils of 20% in valerian, over 80% in Angelica archangelica, 10% in sage, and 10% in peppermint. It prevents mold and spoilage when grown near tomatoes, and it stimulates humus formation. Sage repels cabbage butterfly among cabbages and makes cabbage more tender and digestible. Sage and rosemary are mutually beneficial. When grown with roses, garlic and onions increase the roses' perfume. Onions also repel the rose bug, and roses like compost made with garlic and onion refuse. Parsley grown nearby also aids roses and is good for tomatoes. Strawberry grows well with thyme and borage. Carrots thrive with lettuce and chives and in turn assist the growth of lettuce and leeks. Summer savory is recommended as a border for onions. Yarrow, rosemary and lavender make excellent borders for gardens.

Many plants inhibit the growth of others. Rue and basil dislike each other; so do caraway and fennel. Fennel inhibits beans, tomatoes, and kohlrabi. Chamomile hampers the formation of essential oil in peppermint.[26]

The whole focus is directed toward forming a healthy farm or garden which approximates the teeming diversity of nature. Monocultures are broken by companion planting, mixed crops, wild or carefully planted hedgerows, and an appropriate rotation. Certain herbs are added to grass and clover seed mixtures for pasture to increase diversity and for animal and soil health. Production is not gained at the expense of animal, plant, or soil health and produce must be tasty, nutritious, and have good keeping properties. The farm or garden is treated as a living organism, an interrelated totality.

### The Biodynamic Preparations

At the heart of the biodynamic method sit the "preparations", which are referred to today by numbers. There are two groups of them: two sprays and six compost additives. The sprays are No. 500, cow manure placed in a cow horn and buried for a winter; and No. 501, quartz placed in a now ho-m and buried for a summer. Preparations No. 502—507 are made from yarrow blossoms, camomile blossoms, stinging nettle, oak hark, dandelion flowers, and valerian flowers, respectively, and are added to composts and manures. No. 502 (yarrow) is fermented in deer bladders in the earth for six months during the winter. It helps plants to use the highly diluted sulfur, guides the forces of potassium, and helps draw various trace elements into the soil. No. 503 (camomile) is fermented in the small intestine of cows over the winter in the earth. It governs the access of calcium in plant growth and helps to prevent plant disease. No. 504 (stinging nettle) is buried in humus, two feet deep, insulated by a thin layer of peat moss. This nettle preparation helps regulate sulfur, potassium, calcium, and especially iron in soil and plant. No. 505 (oak bark) is buried in a cow skull over winter. Oak bark contains 77% calcium, an important element in plant disease prevention. No. 506 (dandelion) is buried in a ruminant mesentery. It mediates between silica in the soil and the atmosphere. No. 507 is cold-pressed from valerian flowers as juice. It is used (5-10 drops to 2 gallons of water) as a spray over the base and tops of compost heaps. It attracts and stimulates the propagation of earthworms, and when sprayed on plants helps prevent frost damage. Its high phosphoric content aids in the assimilation of phosphorus needed in plant growth and flowering. An additional spray, No. 508 is made from horsetail (Equisetum arvense), the high silica content of which makes it a good prophylactic agent against fungus and mold.

E. E. Pfeiffer spent many years testing and experiment! with these preparations. He found that the preparations stimulated yeast growth even after being sterilized, pointing to the presence of heat-resistant hormones, important due to the heat in the compost fermentation process. The effect on yeast growth was an extremely high increase, even in dilutions of up to 1:100 million. In practice, Nos. 502-507 are used 2 grams to 15 tons of manure, or in a concentration of 1:75 million. Nos. 500-501 are used in a concentration of 1:25 million. In growth hormones, such dilutions are not at all unusual.[28]

A number of trace elements are increased during the fermentation process, including molybdenum, vanadium, manganese, and titanium, which are essential to the proper growth and nutritive value of plants and bacteria. For example, the action of nitrogen-fixing bacteria need the presence of molybdenum and vanadium. The work of these bacteria is evident in increases of nitrate nitrogen in the preparations: No. 500, 29.3 times; No. 502, 15.8 times; No. 503, 77.5 times; No. 506, 10 times. In compost and manures, the preparations remove the fecal bacteria and those bacteria which break down nitrogen to ammonia or nitrates with a subsequent loss of nitrogen, while nitrogen-fixing bacteria appear. It is interesting to note that when Steiner formulated these preparations, science had yet to discover the necessity of trace elements for plant growth.[29]

**The Preparations and the Cosmic and Earthly Forces**

It is important to remember that the biodynamic farmer seeks to deal with life processes as well as substances. The preparations "are based on the living plant, its contact with the living soil, and its connection with a living cosmos."[30] Just as Steiner chose the preparation substances for their specific qualities, so did he select the animal organs for their role in containing and creating the finished product, and he gave keys toward understanding this function. A full explanation goes far beyond the scope of this exposition, and into the anthroposophical view of zoology.[31] In the animal kingdom, Steiner asserted, there exists a third principle which lifts the mineral and etheric functions of the plant to a higher level. This principle, which he called "astral," still present in the animal organ, adds a new quality to the fermentation process through a kind of "digestion," just as animal manure differs qualitatively from vegetable compost. Thus the animal principle, the principle of consciousness, is necessary to convert the preparation substances into bearers, or carriers, "of the fully-developed processes required by plant and soil."[32]

The preparations harmonize the fermentation of compost and manure, add growth hormones to plant and soil, thus stimulating humus formation and stability. They help the biological processes in the soil to regulate the trace elements and growth hormones, which are themselves very active and very crucial to plant growth. Thus through the preparations the farmer assists nature to balance her life processes for balanced growth conditions. In the available space, I can only state these principles and some of the experimental data. I bring it up because it illustrates Steiner's intuitive insight into the finer processes of nature, and the necessity for a fluid, enlivened thinking to enter into these processes in order to mediate them in farming and gardening.[33]

Preparation 500 (horn-manure) and Preparation 501 (horn-silica) are used as sprays to enhance the polar-opposite forces of earth and cosmos. Preparation 500 is sprayed directly on the soil shortly before or after cultivation to prepare for planting by stimulating soil life. Preparation 501 is sprayed on the leaves of plants when the plant organ that i? to he harvested is beginning to develop and further spraying is done when ripening begins. Horn-manure is applied usually in late afternoon when the damp is beginning to fall, while horn-silica usually is used in the early hours of a sunny morning.[34]

Much statistical data and observations are available or the effects of these two preparations on growth and ripening. They are closely allied with, and support and improve, the earthly (horn-manure) and cosmic (horn-silica) forces. Perhaps the following table constructed by Dr. H. H. Koepf will provide an illustration of this polarity:[35]

Yield and quality under the influence of polar opposite growth factors

| Earthly influences | Cosmic influences |
|---|---|

include among others:

| | |
|---|---|
| soil life; nutrient content of soil; water supply; average atmos-pheric humidity. | light, warmth and other climatic conditions, and their seasonal and daily rhythms. |

vary locally according to:

| | |
|---|---|
| clay, nutrients, humus, lime, and nitrogen content of the soil; | sun; cloudiness; rain; geographical latitude; altitude and degree of nutrient and water holding expo- |
| sure; aspect of land; annual capacity; temperature and pre-cipitation. | weather pattern; silica content of soils, etc. |

normal influences on growth are:

| | |
|---|---|
| high yields; protein and ash content. seed | ripening; flavor; keeping quality; quality. |

one-sided (unbalanced) effects are:

| | |
|---|---|
| lush growth; susceptibility to diseases ter | low yields; penetrating or often bit- |
| and pests; poor keeping quality. fruit; | taste; fibrous woody tissue; hairy |
| | pests and diseases. |

managerial measures for optimal effects are:

| | |
|---|---|
| liberal application of manure post; | use of mature manure and com- |
| and compost treated with biodynamic preparations; sufficient legumes in rotation; compensating for deficiencies; irrigation; mulching. | fertilizing; compensating for deficiencies; suitable spacing of plants; amount of seed used. |
| <u>Use of Preparation 500</u> | <u>Use of Preparation 501</u> |

The cosmic forces of light and warmth, and the planetary influences which affect growth, work an molding or shaping agents, giving form to the plant substances. The atmosphere, mediates these forces, and siliceous substances mediate their activity in soil. At the other pole, the earthly forces provide substance for plants, and are mediated through humus, calcium, water and nitrogen. The interplay between the two poles must be properly balanced to achieve nutritious produce. An overabundance of earthly forces causes high yields, delayed ripening, and a lack of true protein formation with large amounts of free amino acids, amides, or in extreme cares, nitrates. A predominance of cosmic forces can cause premature bolting and seed formation, woodiness and bitterness.[36]

Various soils demonstrate the working of earthly-cosmic forces. Where silica predominates, as in light, sandy areas, plants with good structure and form thrive, such as grasses and cereals. The silica content in grasses is 10-20 times higher than in legumes, for example. Silica in soils is absorbed by plants during water intake. It helps prevent fungal diseases. Silica tends to be found in the more external parts, pointing to its formative quality. Too much nitrogen fertilizer will hamper the work of silica, causing an overly lush, watery, weak growth. At the opposite pole is calcium, usually present in soil as limestone. In contrast to the formative nature of silica, limestone works mainly in plant metabolism and soil biochemistry, and is necessary for the building of stable humus, and for the regulation of mineral nutrients. Here, again, we can see the necessity, for a proper balance between the polarities.[37]

When light is predominant, the cosmic forces outweigh the earthly; when light is lessened, the earthly forces take over. In numerous experiments, scientists have observed that inorganic nitrate fertilizer produces effects similar to an overabundance of earthly forces, or to a lack of light. In other words, heavy nitrate application causes the same typical forms that plants have when grown in shade. These morphological experiments showed that biodynamic preparations in compost can counteract the effects of overabundant earthly forces.[38]

### A Brief Summary

The biodynamic method seeks to build a farm or garden into a living, balanced organism in which soil, plants, animals, man, and cosmos are interrelated. The proper balance achieves a state of equilibrium which reduces the one-sided growth of pestilential insects to the point that poisons become unnecessary.[39] Every effort is made to conserve organic matter in the soil and to build soil fertility to insure plant and animal health. But even more importantly, the biodynamic method demands an active collaboration between the farmer and the kingdoms of nature. It demands that his observation and thinking become continually more subtle and living in order to see the forces of nature as a living, pulsating *whole*—a totality.

The farmer or gardener, through exercises giver by Rudolf Steiner, can undertake to merge his inner and outer life so that his professional development proceeds apace with his self development. The alienation and separation that characterize the modern consciousness thus meets a formidable healing force, a force that acts on two levels: an inner growth and development, and creative relationship with the outer world. Biodynamic agriculture heals the earth through an enhancing of the earthly-cosmic rhythms and forces; the demands this makes upon the farmer can serve as a powerful antidote to the materialistic world-view of our culture. In order properly to mediate the cosmic and earthly forces, the farmer must bring these forces to life in his thinking. Just as new life is born in his soil, plants, and animals, so must new life be born within soul. A new thinking results from this birth, a cognition capable of perception. To refer back to Chapter Four, the Logos becomes active in the soul; life takes on a new meaning as the farmer begins to perceive the forces that pervade nature, and his dally tasks can take on a deep personal significance.[40]

## Practical Aspects

Perhaps we should turn from such "lofty" matters to more earthly concerns, mindful that Steiner sought a marriage between the old peasant intuition and the scientist's precision and analysis in the light of spiritual realities. While the biodynamic method and the philosophy that underlies it afford an opportunity to bring new spiritual forces into the earth, this concern in no way hampers practical farming, but rather assists it. In addition to the ecological benefits of no poisons and no inorganic chemical fertilizers, the increased and stabilized soil fertility improves the farm year by year. The healthy soil and plants produce healthy livestock, which are not stimulated to over-production; as a result, veterinary bills are often much lower than on other farms.

At Ceres Farm, a biodynamic dairy farm in New York of 340 acres, calf losses have been less than 1.5 percent, while the normal rate is 10 percent. The animals have few metabolic or reproductive disorders due in part to the high-quality feed raised there. The farm was begun in 1959 with 5,000 in capital and 100 percent debt. By 1977, the original 220 acres were 340, and the debt was down to ten percent of the total market value of the farm. The milk production has doubled since 1960.
Each cow now produces thirty percent more milk with sixty percent more fat. Net income has increased from 17.7 percent of gross income to 26.5 percent. At the same time, the percentage of feed brought in has decreased from 18.1 percent to 8.4 percent of gross income.

> In conventional farming an increase in net income has generally been achieved by increasing the production by raising the input of feed, fertilizer, or pesticides. The net income on this biodynamic farm was increased by lowering the input of feed bought and by increasing production at the same time. This is a true increase of productivity that has its implications for the farm as a private enterprise, and also for the national economy.[41]

The Zinniker Farm in Elkhorn, Wisconsin is the oldest biodynamic farm in the United States. This farm of 160 acres, a family operation like the Ceres Farm, produces 90-100 bushels of corn per acre, 75 bushels of oats per acre, and 2-2.5 tons or hay per acre. These figures compare quite favorably to all but the most radical chemically-fertilized farms. Here, too, the veterinary bills are very low (in 1974, a total of $241.00 including pregnancy tests) and milk-butterfat figures are good. The on-farm production of feed provides a welcome cushion against constantly fluctuating feed prices and availability. The Zinnikers entertain a constant flow of visitors to their farm, and it has become a Midwestern center for biodynamic conferences and festival celebrations.[42]

## Quality and Quantity

The discussion of "yield" among agriculturalists can give a misleading impression of biodynamic farms and gardens for several reasons. Biodynamic farms eschew the "cash-crop" monoculture of many large-scale farms. While their yields are respectable compared to the chemically fertilized operations, usually average or above average, extremely high yields do not occur. This is due to the emphasis on mixed crops and proper crop rotations. Advantages accrue to the mixed approach in the reduced exposure to market fluctuations, generally a more regional approach to marketing, and of course the environmental benefits in soil conservation and improvement and in the tremendous energy conservation due to the non-usage of chemical fertilizer. The experience of Ceres Farm and many others shows that today's conditions actually make it increasingly important economically to create a farm organism that can produce its own fodder and fertilizers in the form of manure and vegetable composts.[43]

Yield as a criterion of agricultural success points to the need for a balanced perspective—quality should be considered as well. The biodynamic movement has proceeded in the search for quality along several paths: scientific testing of plant and soil, and cooperative marketing with strict quality controls. I mentioned earlier the work of Kolisko, Pfeiffer and others in qualitative and quantitative research.[44] Conventional testing is used to test plant substances, although unconventional questions are asked. In addition to chemical analysis, physiological tests are used to observe the changes themselves as a manifestation of life processes. For example, a pathological test uses pathogenic organisms to measure decomposition in harvested plants. Goethean morphology observes the form of plants—leaf development, root growth, fruit formation—to evaluate field experiments. Finally, scientists developed two methods to create pictures to test the life forces in plants and soils, although these have not been generally recognized by the conventional scientific community, even though their usefulness has been proven. The reason seems to stem from the fact that reading the pictures requires diligent practice and is somewhat difficult to master. A "picture" of quality does not lend itself easily to quantification.

E. E. Pfeiffer developed the crystallization method, in which copper chloride solution is allowed to crystallize on a flat plate, forming a haphazard crystal pattern. When a solution of plant sap or other organic material is dropped onto the plate, "the crystals form an organized picture that is specific for that particular solution."[45] This test involves a similar process as that when a plant organizes substances from air and soil into its specific substance and shape; thus the plant sap organizes the copper chloride crystals that were previously arranged in no ordered pattern. A significant body of research has arisen out of this test.[46]

Lily Kolisko invented the capillary dynamolysis test. An organic juice is applied to a filter paper to which a solution of metal salt is added. The resulting pattern of shapes and colors is then "read". This test, later modified by Pfeiffer into the chromatogram test, is used to test soils, composts, and plants. I may remark parenthetically that both tests are used extensively in anthroposophical medicine to test the blood for the presence of disease and in the selection of plants and harvesting times for herbs to be used for medicines.[47]

In Germany, the Demeter Cooperative began work in 1928. After changing its name to the Demeter Trading Association, (Demeterwirtschaftsbund) it operated until the Nazis banned the organization in 1941. The Demeter has functioned again since 1954. Today it includes producers, processors, distributors and consumers. Regular meetings of representatives from the various European nations set uniform standards and guide,lines. To use the trademark, the member must pass quality tests, have used biodynamic methods for not less than two years, and of course use no chemicals as fertilizers or poisons. Even bought-in organic fertilizers must be limited. These rules apply both to plants and livestock. For the past seven years, there has been a Demeter certification available for American farmers. It conducts regular on-site inspections, grades the farm according to how it meets the rigorous Demeter standards, and then allows the use of the registered Demeter trademarks. The Demeter certification is respected internationally as a sign of the highest quality produce.[48]

## Concluding Comments

The picture that emerges from the foregoing considerations, when viewed in the light of the previous chapters, reveals an agricultural method that makes quite heavy demands upon its practitioners. Biodynamics requires a fluid, strong thinking capable of dealing with the complex interactions of earthly-cosmic rhythms and forces. The farmer can enliven his thinking and enhance his perceptions through exercises given for this purpose by Rudolf Steiner. Steiner hoped that a fruitful relationship would ensue between biodynamic farmers and anthroposophical scientists, that science would be enriched and grounded through the instinctive certainty and common sense of those who worked the earth, while science would bring its acute observation and clarity to farming. The image of the farmer, then, is of one who actively seeks to work with life processes in soil, plant, and animal, to harmonize the cosmic-earthly Forces, and to create a living farm organism that exists as a harmonious part of the landscape. The farmer must be part scientist, peasant, and artist.

Many of the judgments the farmer must make are aesthetic in their nature. Questions of balance and harmony begin to emerge from the purely scientific or agricultural considerations. Whether it is reading the results of chromatograms or crystallization tests, laying out a field with hedgerows, planning a herb garden, selecting a heifer to keep as a replacement cow, choosing produce for the market, all these things require some sense for beauty, for artistic vision of form, proportion,,etc. No matter how many college courses one has taken in cattle-judging, landscaping, or horticulture, the actual decision, if it adds to the beauty of farm or herd, has an aesthetic quality. No one can tell you, as an intellectual rule, when a thyme or cabbage seed is ready to harvest when it has achieved just that maximum concentration of essential oils that ensures excellent keeping and germination.

In tracing Steiner's concept of the evolution of consciousness, we saw the unity of art, science, and religion of the ancient world grow increasingly fragmented. Steiner attempted, through his spiritual science, to provide a basis for a renewal of culture—a re-unification. The work that has resulted from his efforts, in agriculture, medicine, or education, demands a large degree of self-development, that one's professional work becomes in a certain sense artistic, and this requires deep reverence, wonder, and devotion toward life. Living begins to become an art, and work becomes quite the opposite of the boring, compartmentalized, over-specialized drudgery that occupies many modern human beings. The biodynamic farmer or gardener thus approaches his work with the same feelings that ancient mankind reserved for the mystery-centers, the temple that was the earthly embodiment of the old unity of culture. While he may not precede his plowing or sowing with a ritual or incantation, the deeply reverent attitude is there.[49] Steiner felt that the feelings of reverence, devotion, and wonder are the prerequisites for higher cognition and perception, and we have seen that the creative collaboration of the farmer with nature necessitates a merging of professional and self-development. How can thinking and feeling be enlivened to the point where they assist in self-development, and in dealing with the living world of nature? We must now turn to the curriculum of Steiner schools to explore ways that life-affirmative values can be nurtured.

**Footnotes for Chapter Six**

1. Barry Commoner, *The Closing Circle: Nature, Man and Technology* (New York: Alfred A. Knopf, 1972) 7.
2. Rudolf Steiner, *Agriculture* (London: Biodynamic Agricultural Association, 1974).
3. E.E. Pfeiffer, *Biodynamics: Three Introductory Articles,* Springfield, Ill.: Biodynamic Farming and Gardening Association, 1948, 1956) 29; Guenther Wachsmuth, *The Life and Work of Rudolf Steiner* (New York: Whittier Books, 1955) 546.
4. Johannes Hemleben, *Rudolf Steiner: A Documentary Biography* (East Grinstead, Sussex: Henry Goulden Ltd., 1973) l; Wachsmuth, Ibid., 512-571, for Steiner's activities in 1924.
5. E.E. Pfeiffer, "New Directions in Agriculture," *Golden Blade* (1958) 118, 121.
6. C.A. Meir, "What Is A Farm?" in A.C. Harwood, ed., *The Faithful Thinker* (London: Hodder and Stoughton, 1961) 223; Koepf, Pettersson, and Schaumann, *Biodynamic Agriculture: An Introduction* (New York: 1976) 16-17; and Pfeiffer, "New Directions in Agriculture," 119. For an account of biodynamics in the United States, see Elizabeth Speiden Gregg, "The Early Days of Biodynamics in America," *Biodynamics* 119(Summer, 1976) 25-39; 120(fall, 1976) 7-21; and 121(Spring,1977) 16-23. Also see Helen Philbrick, "Biodynamics in Recent Years," 122(Summer, 1977) 20-23.
7. George Adams, "Space and Counter-Space," in Harwood, ed., *The Faithful Thinker*, 126-127.
8. C.A.Meir, "What Is A Farm?" 225. See also Guenther Wachsmuth, *The Etheric Formative Forces in Cosmos, Earth and Man* (London: 1932) ; Agnes Fyfe, *The Signature of Mercury in the Plant* (Stuttgart: Verlag Freies Geistesleben, 1972); George Adams, Ibid., 125-144; George Adams and Olive Whicher, *The Plant Between Sun and Earth* (Clent, Worcester: Goethean Science Foundation, 1952); and Olive Whicher, *Projective Geometry* (London: 1972).
9. H.H. Koepf, "What Is Bio-dynamic Agriculture?" (Springfield, Ill.: Biodynamic Farming and Gardening Association, 1970) 10; C.A. Meir, "What Is A Farm?," 225-226. See the thorough discussion of levity in Ernst Lehrs, *Man or Matter* 2nd ed. (London: Faber and Faber,1958) 167-192. For some examples of Goethean science and research, see note 9 and also Jochen Bockemühl, "Toward the Characterization of Plant Processes and the Evaluation of Food Quality, Using the Radish as an Example," *Biodynamics,* 118(Spring, 1976) 12-28; Maria Thun, "Nine Years Observation of Cosmic Influences," *Biodynamics*, 115(Summer, 1975) 14-22; and her *Work on the Land and the Constellations,* (East Grinstead, Sussex: Lanthorn Press, 1977). Other examples of the anthroposophical work in the sciences: E.E. Pfeiffer, "A Dynamic Concept of the Weather," *Biodynamics*, 4(Spring, 1946) 1-6; Michael Wilson, "Colour, Science, and Thinking, in Harwood, ed., *The Faithful Thinker*, 141-152; Heinz Grotzke, "Rudolf Steiner s Impulse To Herbology," *Biodynamics,* 90(Spring, 1969) 20-29; T.V. Lai, "Phosphorus and Potassium Uptake by Plants Relating to Moon Phases," *Biodynamics,* 119(Spring,1976) 1-14; George Adams, "Potentisation and the Peripheral Forces of Nature," *Golden Blade* (1966) 23-40; Wilhelm Pelikan and Georg Unger, "The Activity of Potentized Substances," *British Homeopathic*

*Journal,* 60,4(October, 1971) 223-265; the latter two articles are very helpful in understanding the action of substances in very high dilutions, such as the biodynamic preparations. Finally, C.P.J. Lievegoed, *The Working of the Planets and the Life Processes in Man and Earth* (Clent, Worcester: Experimental Circle of Anthroposophical Farmers and Gardeners, 1951) and Lily and Eugen Kolisko, *Agriculture of Tomorrow* (Edge, Gloucester: Kolisko Archive, 1946) has excellent information on the working of the preparations, experiments, etc. A beautiful Goetheanistic study is Gerbert Grohmann, *The Plant: A Guide to Understanding its Nature* (London: 1974); also J. Bockemühl, *Toward a Phenomenology of the Etheric World* (New York: 1988).

10. Rudolf Steiner, *Agriculture*, 65. See also Hilmar Moore, "The Living Earth," *Biodynamics*, 170(Spring 1989) 34-43.

11. Rudolf Steiner, *A Modern Art of Education*, 3rd ed. rev.(London: 1972) 221.

12. Koepf, Pettersson, and Schaumann, *Bio-Dynamic Agriculture*, 27. For Steiner's self-development exercises and meditations see *Knowledge of the Higher Worlds and its Attainment,* 3rd ed. (New York: 1947); *A Road to Self Knowledge and the Threshold of the Spiritual World,* 3rd ed. (London: 1975) and chapter five of *An outline of Occult Science* New York: 1972) 246-347.

13. Barry Commoner, *The Closing Circle*, 181.

14. Koepf, Pettersson, and Schaumann, *Bio-dynamic Agriculture*, 27.

15. I am indebted to a conversation with Dr. H.H. Koepf on 28 August 1976 at the Biodynamic Farming and Gardening Conference, Spring Valley, New York. See his discussion in Ibid., 27-29.

16. Pfeiffer, *Biodynamics*, 30.

17. Lynn White Jr., *Medieval Technology and Social Change* (London: Oxford University Press, 1962) 39.

18. Rudolf Steiner, *The Tension Between East and West* (London: Hodder and Stoughton, 1963) 114.

19. Ibid.

20. Steiner, *Agriculture*, 63, 64.

21. Ibid., 84-85.

22. Koepf, *What Is Biodynamic Agriculture?*, It.

23. Koepf, Pettersson, and Schaumann, *Biodynamic Agriculture*, 29, 31. I am indebted to John Davy of Emerson College, Sussex, for the concept of the relationship between professional development and self development. Interview, Detroit, Michigan, 15 March 1977.

24. Koepf, *What Is Biodynamic Agriculture?,* 12; Koepf, et al, *Bio-Dynamic Agriculture*, 30-31. The quote is from C.A. Meir, "What is A Farm?," 227.

25. E.E. Pfeiffer, *Biodynamics*, 3, 5, 32-33, This discussion, is taken from Helen Philbrick and Richard Gregg, *Companion Plants and How To Use Them* (Old Greenwich, Conn.:-Devin- Adair, 1966) and Koepf et al, *Bio-Dynamic Agriculture*, 232-234.

26. Ibid.

27. See Pfeiffer, *Biodynamics*, 11-26; Koepf et al, 206-224.

28. The data are from Pfeiffer. Ibid ; Lily Kolisko, *Agriculture of Tomorrow Preparations* (Edge, Glos.: Kolisko Archive, and E.E Pfeiffer, *Soil Fertility, Renewal;, and Preservation,* (E. Grinstead: Lanthorn Press, 1983) 130-147..

29. Pfeiffer, *Biodynamics*, I1-19; and Heinz Grotzke, "Rudolf Steiner's Impulse to Herbology," *Biodynamics*, 90(Spring, 1969) 27-29.

30. C.A. Meir, "What Is A Farm?" 226.

31. Ibid., 227. For the anthroposophical view on zoology see Hermann Poppelbaum, *Man and Animal: Their Essential Difference*, 2nd ed. (London: 1960); and his *A New Zoology* (Dornach: Philosophic-Anthroposophic Press, 1961); and Eugen Kolisko *Zoology for Everyone* (Edge, Glos.: Kolisko Archive, 1947).

32. C.A. Meir, Ibid., 227-228.

33. Koepf, Pettersson, and Schaumann, *Bio-dynamic Agriculture*, 205; see also H.H. Koepf, *The Biodynamic Farm* (New York: 1989) for a wide variety of experimental data on many aspects of biodynamic farm management.

34. Koepf, *What is Biodynamic Agriculture?*, 14-15; Ibid., ?13-21.

35. The table is from Koepf et al, Ibid., 209.

36. Koepf, *What Is Biodynamic Agriculture?*, 27-28.

37. Koepf, Pettersson, and Schaumann, *Bio-dynamic Agriculture*, 183-186.
38. Ibid., 363-368.
39. Pfeiffer, *Biodynamics*, 10.
40. See Hartmut von Jeetze, "In Defense of Old-Fashioned Training," *Biodynamics*, 122(Spring, 1977) and 123(Summer, 1977) and his "Bio-dynamic Relations Between Man and the Land," *Biodynamics*, 115(Summer, 1975) 10-15. See also, Philip Very, "A Comparison of Individuals Attracted to Various Agricultural Methodologies," *Biodynamics*, 91(Fall, 1969) 12-22.
41. Koepf, Pettersson, and Schaumann, *Bio-dynamic Agriculture*, 249-251.
42. Ibid., 255-256. See also Hilmar Moore, "Ruth Zinniker—A Life in Biodynamics," *Biodynamics*, 170(Spring 1989) 15-21.
43. Ibid., 39, 70-93; and Koepf, *What Is Bio-dynamic Agriculture?*, 34-35.
44. See notes 8 and 9.
45. Koepf, Pettersson, and Schaumann, *Bio-dynamic Agriculture*, 356-362. The quote is from 362.
46. Ibid..
47. Ibid., 362-363; Stewart C. Easton, *Man and World in the Light of Anthroposophy*, (New York: 1975; 476-477. Koepf, Pettersson, and Schaumann, *Biodynamic Agriculture*, 391-396. For more on the Demeter certification organization, see H.H. Koepf, *The Biodynamic Farm*, 186-188, 231-236. Koepf not only describes the certification qualifications and process, but he also gives many examples of the quality of biodynamic produce which have been demonstrated by various experiments.
48. Write Demeter Association, Inc. for certification of biodynamic produce. Address: 4214 National Ave., Burbank, CA 91505.
49. When asked why he farmed biodynamically, Richard Zinniker replied, "I guess you could say it's my religion." Conversation, 12 November 1975 at Zinniker Farm, Elkhorn, Wisconsin.
50 . See note 12.

# CHAPTER SEVEN

## Waldorf Education and Ecological Values

*We ought not to be satisfied, for instance, with presenting a plant, a seed, a flower to the child merely as it can be perceived with the senses. Everything should become a parable of the spiritual. In a grain of corn there is far more than meets the eye. There is a whole new plant invisible within it. That such a thing as a seed has more within it than can be perceived with the senses, this the child must grasp in a living way with his feeling and imagination.*

*Rudolf Steiner*[1]

*The first essential thing is to awaken in them a feeling for the forces of growth, for the eternally creative forces of Nature. The next step is to awaken in them a sense of responsibility toward these forces of growth, towards the health of the soil, of plants, of animals and of men, and also an inner sense of satisfaction in progressing towards this goal. This in turn becomes a compensation for the modesty of the livelihood earned. Those who cannot develop these ethical qualities will never become good farmers or colonizers, and they will hardly ever become constructive members of the social organism.*

*E. E. Pfeiffer*[2]

### A Historical Note

A characteristic of Rudolf Steiner's practical work is that he had to be asked to consider a problem before he began to deal with it, a good example being the Agricultural Course. A similar series of events resulted in the Waldorf School in Stuttgart, the first Steiner school, which opened 7 September, 1919. Steiner had given a lecture on education in 1908 called "The Education of the Child in the Light of Anthroposophy", in response to many requests.[3] It was not until after World War I ended, with the ensuing chaos in Germany, that anyone resolved to press Steiner for more pedagogical information. In 1919, Emil Molt, director of the Waldorf Astoria Company in Stuttgart and a student of Steiner's work, concerned himself with providing proper education for his workers' children. Molt had read "The Education of the Child" and wondered how to begin a school. Then the workers' enthusiasm for Steiner's ideas after Steiner spoke to them led Molt to forge ahead with plans for the school. Steiner himself selected the teachers, and on 21 August, 1919, he began two series of parallel lectures and a seminar. The lecture courses today are published as *Practical Course for Teachers* and *The Study of Man*, and the record of the seminar as *Discussions with Teachers*. The meetings ended on 5 September, and the Waldorf School opened on 7 September 1919.[4] These courses still form the foundation of today's training courses on Waldorf pedagogy.

During the next five years, Steiner attended many of the teachers' meetings at the first Waldorf school. The first teachers all had acquired previously a basic knowledge of Steiner's philosophy, so that the intense two-week session of lectures and seminar was a culmination of their preparation, an application of the insights of spiritual science to education. From his lectures and indications to these teachers, and several other cycles of his lectures, has arisen the Waldorf pedagogy and its curriculum.[5] In the years since 1919, Waldorf schools, named after the Stuttgart Waldorf School and also called Steiner schools, have been established in Europe, Scandinavia, Great Britain, Canada, the United States, Brazil, Mexico, New Zealand, Australia, and Africa. The situation today finds a rapidly expanding movement in which the available teaching positions exceed the supply of trained teachers.

In the second chapter, I described some of the basic features of Steiner education. This essay will not go into the foundations of Steiner's pedagogy in great detail, but rather will attempt to discuss the aspects of the curriculum at different age levels that can be helpful in forming the attitudes necessary for an understanding of biodynamics and ecology.

## The Foundation of Waldorf Education

Rudolf Steiner emphasized the metamorphic quality of human life. What happens at an earlier time in life affects one for many years afterward, often in unexpected ways. To begin to understand life, one must peer beneath its surface:

> Life in its entirety is like a plant. The plant contains not only what it offers to external life; it also holds a future state within its hidden depths. One who has before him a plant only just in leaf, knows very well that after some time there will be flowers and fruit on the leaf-bearing stem. In its hidden depths the plant already contains the flowers and fruit in embryo; yet by mere investigation of what the plant now offers to external vision, how should one ever tell what these new organs will look like? This can only be told by one who has learnt to know the very nature and being of the plant.
>
> So, too, the whole of human life contains within it the germs of its own future; but if we are to tell anything about this future, we must first penetrate into the hidden nature of the human being; And this our age is little inclined to do.[6]

Steiner developed his concept of the rhythmically successive periods of development and his threefold model of the psychosomatic relationships in an attempt to provide such an understanding of the nature of humanity. The educational impulse that resulted from these considerations is based on a precise knowledge of the human being, of the threefold nature of body, soul, and spirit seen in the three bodily systems: the head-sensory-nerve system; of the rhythmic-circulatory system of heart and lungs; and of the metabolic-limb system. In soul activity, Steiner held that thinking corresponds to the nerve-sense system; will finds its somatic basis in the metabolic-limb system; and feeling—the mediator between the polarity of conscious thinking and unconscious will—corresponds to the bodily rhythmic system. One can see the head-nerve-sensory system as the bodily basis of spiritual activity—thinking—while the metabolic-limb system supports the more purely physical activity. Thus the rhythmic system is the physical foundation of the soul; it mediates between the nerve-sense and metabolic-limb polarity just as the soul forms the mediator between spirit and body.

In addition to the threefold psychosomatic model, Steiner also gave a fourfold model, one for which part of the preceding chapter prepared us to consider here. This perspective of human nature relates the human being to the kingdoms of nature and postulates a series of four bodies; by *body* Steiner meant that which gives a being shape or form.[7]

He wrote that we have a *physical body* in common with the whole of the mineral kingdom. This body is subject to the same laws of physical existence, and is built up of the same substances and forces, as the whole of that world which is commonly called lifeless.[8] Animals, plants and human beings all have such a mineral body.

Just as we saw in biodynamics, there is an essential principle of life which enlivens the mineral constituency of the plant, shaping and forming it into a living entity, and providing the processes of reproduction, the inner movement of saps and fluids, heredity, and growth. The human being shares this body or principle with the plant kingdom and the animal kingdom. Steiner stated that it is therefore the builder and molder of the physical body, its inhabitant and architect. The physical body may even be spoken of as an image or expression of this life-body.[9] The life-body, formative body, or *etheric body*, as Steiner variously called it, thus is a second entity within the human organism.

The third member of the human being Steiner called the *astral body* or sentient body. He described it as the body of consciousness:

> It is the vehicle of pain and pleasure, of impulse, craving, passion, and the like—all of which are absent in a creature consisting only of physical and etheric bodies. These things may all be included in the term: sentient feeling or sensation. The plant has no sensation. If in our time some learned men, seeing that plants will respond by movement or in some other way to external stimulus, conclude that plants have a certain power of sensation, they only show their ignorance of what sensation is. The point is not whether the creature responds to an external stimulus, but whether the stimulus is reflected in an *inner* process—as pain or pleasure, impulse, desire or the like.[10]

Humanity possesses a sentient body only in common with the animals; plants do not have within them a sentient body. This is the reason for using animal organs in fermenting some biodynamic preparations, in order to sheathe the herb with the animal essence provided by the organ.

The fourth constituent of the human being, Steiner held, humanity shares with no other living creature. This is the ego, or the human I, a name essentially different from all other names. As he put it:

> No one can use this name to designate another. Each human being can only call himself 'I'; the name 'I' can never reach my ear as a designation of myself. In designating himself as 'I', humanity has to name himself within himself. A being who can say 'I' to himself is a world within himself. Those religions which are founded on spiritual knowledge have always had a feeling for this truth. Hence they have said: With the 'I', the 'God'—who in the lower creatures reveals himself only from without, in the phenomena of the surrounding world—begins to speak from

> within. The vehicle of this faculty of saying 'I', of the Ego-faculty, is the 'Body of the Ego', the fourth member of the human being.[11]

Steiner emphasized that these bodies are not simply increasingly finer substances of the material world, but rather are figures of active forces, not bound by material considerations. It is possible to view the polarity of the will activity of the soul and the limb-metabolic system on the one hand and the nerve-sensory system and thinking on the other hand, through the perspective of these four members. Thus the physical-etheric lies on the life pole, and the astral-ego (or soul-spiritual) exists on the consciousness pole of being. In the interaction between the metabolic processes of the life pole and the destructive catabolic process as the soul-spiritual pole works into the lower members, arises human consciousness.[12] I must add here that Steiner saw the astral body of the plant as outside its physical-etheric body. It thus works around the physical-etheric, shaping the growth forces from without. The astral body does penetrate into the life-forces at one point, however: *when the plant blooms*. The color of the blossoms, Steiner said, reveals the kiss of the astral forces. That these forces occupy a polar relation to the life forces can be seen in the fact that an annual plant forms seed and dies or goes into dormancy soon after blooming.

It is from these three perspectives—the threefold, and fourfold models and the polarities therein—that Steiner developed his pedagogical work. He detailed for each of the seven-year periods of development, and the gradations within them, what exact changes these constituents of the human being undergo.

> So for the art of education it is a knowledge of the members of man's being and of their several developments which is important. We must know on what part of the human being we have especially to work at a certain age, and how we can work upon it in the proper way.[13]

On this foundation he wrought the Waldorf curriculum. Like the principles of biodynamic agriculture, the Waldorf methods originated in Steiner's deep insight into the world of humanity and nature, and his ability to discover the objective truths of the inner life and its proper relation to the world and cosmos.

### The Main Lesson and the Class Teacher

The main lesson and the class-teacher are two of the most notable innovations in Steiner's pedagogy. He felt that the period of a child's development from the change of teeth to puberty, in which the feeling life is so pronounced, places great demands on teacher and school. The feeling activity of the soul, with its somatic basis in the rhythmic system of heart and lung, makes it advisable to create a strong rhythmic element in the school. The main lesson, which occupies the first two hours of each morning, permits sufficient time to create a mood, so important for feeling and art. It allows the teacher really to build up a subject without an immediate switch to another subject at the end of fifty minutes. In the main lesson, subjects are taken in blocks of time, so that each day the children can be immersed in the material for three, four, or even five weeks. In this manner, a rhythm can arise; children can settle down and become open to the subject.

Steiner held that the main lesson could be used to enhance a child's concentration and to enable a deep penetration of a subject.[14]

The class teacher represents an additional recognition of the importance of feeling for this age group. If at all possible, the same teacher accompanies the children from the first through eight grades, and teaches all the main lesson subjects. In this manner the teacher comes to know the children intimately, and the resulting continuity provides a sense of stability for the students and an opportunity for clear insight into each child's development for the teacher. In addition, the teacher forms a close connection with the parents. The class teacher faces a tremendous challenge both in the mastery of new materials and from the stages of child development that change each year. Yet as Werner Glas points out:

> The interrelationship between subject matter taught in earlier and later years adds unique possibilities to his task. He knows what the children have experienced, he can allow a theme to be forgotten and reawakened in another form at a later date, and he can throw many bridges between disciplines which are usually separated academically. In this way he works toward the unification of experience and a sense of relatedness in knowledge, with the image of humanity as its coordinating point.[15]

I have traveled extensively to many schools in the past few years. I have found that some schools are considering modifying the seventh and eighth grades away from a strict following of the class teacher tradition, in favor of using more specialist teachers in some subjects in which the class teacher may feel less well trained. This change does not reflect adversely on the class teacher principle; to the contrary, it merely shows that there is a growing recognition that children are maturing more rapidly today, that the curriculum in the upper grades is quite demanding, and that a teacher who is excellent with the younger children may not have the same gifts for older children and vice versa.

## The Waldorf Curriculum

Before we begin to explore some aspects of Waldorf teaching, I will give a brief outline of the curriculum as it is found, with some variations, in most Waldorf schools. The variations occur because each Steiner school, although they rest on the same philosophical foundation and seek to deal with the ideals that lie in the universal development of humanity, must also take into account the national culture and the needs of their students. In addition, each school takes on its own special character since it is an entirely independent enterprise, resting wholly on the initiatives of teachers and parents and whatever local support it may find.[16] Within this diversity, a similar curriculum is followed; the special character emerges from the individuality of the teachers, pupils, and parents.

A typical curriculum for the main lesson in grades 1-8 is as follows:[17]

MAIN LESSON SUBJECTS

*Grade I:* Large and small letters are learned thoroughly as forms and as sounds.

Reading is taught as a by-product of the relative perfection of writing. Writing, in turn, is evolved from painting and drawing. Simple spelling . . . Numbers from 1 to 100. Elements of addition, subtraction, multiplication and division. . . Fairy tales and nature stories are told and retold or dramatized by the class. These, together with poems and stories coming from daily life form the material for considerable writing.

*Grade II:* Reading. Spelling. First elements of grammar, dictation, simple composition ... Arithmetic, including the simpler multiplication tables . . . Fables and legends, nature lore, and acquaintance with natural home environment.

*Grade III:* Reading. Spelling. Compositions and grammar. Cursive writing begun .... Through stories, trips, nature walks, study of farming, housing, and other activities, the sense of practical life is fostered . . . Arithmetic, including higher tables and learning of measures . . . Old Testament stories.

*Grade IV:* Reading. Spelling. Composition, including letter writing. Grammar . . . Arithmetic, including fractions . . . Map-making, Long Island geography . . . Norse myths and sagas . . . Introduction to zoology.

*Grade V:* Composition. Grammar. Spelling. Reading . . . Arithmetic, including decimals and calculations of area . . . Geography . . . Ancient history through Greece (Indian, Persian, Egyptian) . . . Botany.

*Grade VI:* Composition, including business letters, and further development of the feeling for style. Grammar. Spelling. Literature dealing, among other things, with the realm of folk-lore and supplementing historical studies . . . Arithmetic, including an introduction to practical business mathematics: interest, percentage and discount. Geometry . . . Geography . . . Roman and Medieval history . . . Physics: acoustics, optics and heat, mineralogy.

*Grade VII:* Composition, including poetry. Grammar. Spelling. Literature . . . Arithmetic, algebra, and geometry . . . Geography . . . Renaissance, Reformation and the Age of Discovery. American history . . . Astronomy. Physics: electricity, magnetism. Inorganic Chemistry. Physiology.

*Grade VIII:* Composition, including business English. Grammar. Spelling. Literature, especially the epic and dramatic forms . . . Arithmetic, algebra and geometry . . . World geography . . . History from the 17th century to the present. American history . . . Physics. Organic chemistry. Physiology.

SPECIAL SUBJECTS:

French, German, Music, Eurythmy, Art, Handwork, Woodwork and Physical Education are all taught from the first through eighth grades.

These studies are carried on in the high school in quite a different fashion.

The main lesson teacher, with whom the children grow and learn, gives way to the specialist in social studies, literature, art history, the natural sciences and mathematics. The psychologist Susanne K. Langer wrote: "The primary function of art is to objectify feeling so that we can contemplate and understand it. It is the formulation of so-called inward exper ience, the inner life, that is impossible to achieve by discursive thought."[18] She asserted that it is the artistic imagination "which is the medium of self knowledge and insight into all phases of life and mind."[19]

Where the emphasis in the earlier period was on an artistic approach to the subject matter, so that the child can apprehend through an active feeling life, and so that a vibrant world of images and observations can be built, in the high school the approach becomes more intellectual. For Rudolf Steiner, the age of adolescence marks the birth of the intellect. Just as a healthy *feeling* for what is true and good was nourished and nurtured during the first seven years of school, so in the high school a healthy *judgment* a strong and living *thinking* is the goal. It is out of the healthy artistic imagination nurtured in the first eight school-years that a strong thinking with clarity and judgment can be born.

As Stewart C. Easton characterized the high school years:

> At this time the young person becomes ready for numerous changes in his teaching to correspond to the numerous changes that are taking place in his physical and soul-spiritual being. At the change of teeth . . ., those forces that had been hitherto used for creating and thrusting forth the teeth themselves, could be used for building the conscious memory; now they are freed for intellectual thinking, and thinking becomes possible in a way that would have been harmful if it had been systematically trained before.

Easton continued to say:

> Whereas during elementary school it was the rhythmic system that predominated and teaching had to be imaginative and artistic, the adolescent learns above all through observation and thinking, and by the conscious use of his senses. The whole external world now forces itself upon his attention, and for the first time he really becomes "awake."[20]

Several areas of the Waldorf curriculum, particularly language arts, social studies, and the sciences, seem to offer means whereby certain attitudes, and a certain *quality* of thinking about the world and one's place in it, can be fostered. Nevertheless, while the foundation of Waldorf schools rest on Steiner's work, absolutely no attempt is made to inculcate any ideology in the children; rather, the schools try to educate in the Latin meaning of the word, to "draw forth" what dwells within each child "so that it may bear fruit in outer life." Frans Carlgren wrote that

"The task of the teacher is not to infringe on the pupil's self, or 'I', but rather to help the instrument, the body and soul to develop in such a way that the individuality, the spirit, can eventually live freely within it.[21]

I will discuss the actual gardening courses in Waldorf schools in the following chapter.

### Language Arts

The story curriculum of the language arts program leads into the study of history. In Chapter II, I mentioned the progression of these stories and how they parallel the child's phases of development and also the evolution of human consciousness Just as humankind moved from fairy tales to folk tales, to myths and legends, and finally to recorded history, so the Waldorf curriculum follows the same path in the grade schools. The fairy tale takes place "once upon a time" or "everywhere and nowhere". Dragons, witches, gallant princes and beautiful princesses, wise old kings and wicked sorcerers inhabit the stories of many cultures. With folk tales and legends—the latter often selected from the Golden Legend and other stories of saints—the story curriculum leaves the more universal realm and takes a step nearer to earth, so to speak. The fairy tales' "long ago and far away" becomes specific places in the world, which have to be described or characterized by the teacher for the second graders. Animal fables such as those of Aesop or the Indian *Jakatas* are also told. In the third grade begins the first in a series of mythological studies: the third grade uses Old Testament stories, the fourth grade delves into the tempestuous Norse world of giants, trolls, dwarves and gods; and the fifth grade takes Greek mythology.

As I wrote in Chapter V, Steiner held that fairy tales were formulated by great initiates in order to present aspects of spiritual wisdom in picture form, especially those that preserve the old stories in their most original form, such as the Grimm collection. He said that children, with their pictorial, imaginative minds, are quite close to the consciousness of more ancient peoples; thus the world of the fairy tale is one in which they feel at home. By a careful selection from the vast supply of tales, the teacher can tell just those he feels are right for his class. In fact, stories are often found that relate to specific children and their problems. Experience has shown that a story chosen for such a purpose, or created by the teacher, can have a very therapeutic effect, even though no direct mention is made of the child in question. Yet the child receives help in her inner battles, a certain self-recognition that leads to a beneficial healing of a difficult situation.[22] Steiner said that moral character can be taught through fairy stories, not at all in an intellectual, abstract, or didactic way, but simply by allowing the children to live in the powerful images. The teacher seldom reads to the children, at least not in the early grades, but rather tells the story—the way they have been transmitted for many centuries, and the art of which the teacher has spent many hours in perfecting. The folk tales, fables, and legends often take the second grade to quite specific times and places, and introduce the first contact with national character, as in the French Renard the Fox, or in the Irish stories, or in Coyote in native American stories. The fables offer insight into the animal world and imaginatively portray human strengths and weaknesses.

The mythologies afford other opportunities for teaching. The creation myths of the Hebrews, Norse, and Greeks offer striking contrasts. In the myths are many beautiful pictures of nature; Narcissus and Echo, the great Wind Giants and Frost Giants, Zeus' thunderbolts, the various creation myths, the dwarves and gnomes in the dark earth; Persephone and Demeter, and Dionysus come easily to mind. By means of these vibrant images the children are led carefully and gradually into the world. As the children meet these great characters, they also can encounter quite specific plants, animals, the seasonal rhythms, and the elements.

Just as the fairy tales preserved much of the Aristotelian folk-wisdom, so do the myths contain much nature-lore. The children's souls are swathed in a rich tapestry of pictures that contain many levels of meaning and that can provide a lifetime of pleasure and nourishment. And the children have imbibed the ancient atmosphere in which mankind felt the work of the spirit everywhere in nature.

**History**

In the fifth grade, the children begin to explore "the qualitative differences of various cultures."[23] First comes the ancient Indian culture with its huge selection of myths. Perhaps the children already had encountered Indian fables in a previous year. Now they meet stories of Krishna, Rama, and of course, the Buddha. Next, they come to a sharp contrast to the Indian world with the Persian story of Zarathustra, the founder of agriculture, and his role of leading humankind between the powers of light and darkness. Each person must choose whether to follow Ahura Mazda or to fall into the service of Ahriman. Then the culture of Babylon is studied, a culture which influenced both the Hebrews and the Greeks. Particularly important is the epic poem of Gilgamesh, his confrontation with the goddess Ishtar, his deep friendship with Enkidu and the moving death of Enkidu.[24]

With the consideration of Egypt, the first contact with history begins. As Werner Glas describes it:

> When the teacher approaches Egypt, mythology recedes and historical narration comes to the foreground. In a strange way Egypt seems much closer to us. Experience in the American Waldorf schools suggests that children have a deep-seated fascination for things Egyptian, and that they are singularly responsive to the details of a culture which, in actuality, is totally unlike the one we live in. The focus of Egypt was not happiness, but death. Any historical treatment of things Egyptian inevitably leads to discussions of death.[25]

Most Egyptian myths concern the world of the dead, and also "the appearance, disappearance, and reappearance of the sun." Other examples are the earthly death of Osiris, and Hermes accompanying the souls to their judgment before Osiris. The extensive background in mythology assists the children to deal with a culture in many ways dissimilar to their own. In the study of Egypt, emphasis can be placed on the temple life, construction of the pyramids, mummification, and not least on agriculture, animal husbandry, and the Nile irrigation system. The teacher describes the life of an ordinary Egyptian during a whole day's activities.[26]

A striking contrast to the long reign of Egypt in the fertile Nile valley comes with the turbulent history of Greece, a contrast that is seen even in the geography of the two areas. The central culture of Egypt that surrounded the god-like

Pharaoh could hardly be more different than the many Greek city-states, separated by rivers, mountains and the sea, and crisscrossed by trade routes between east, west, north, and south. Many peoples met in Greece and, for all their concern for order, there was constant ferment and clashes, unlike the hundreds of years of peace in Egypt. Suitable Greek myths pave the way for the teacher to build a vivid picture of the unity of Greek civilization, a culture that demonstrates an unerring sense for the wholeness of things. The historian H.D.F. Kitto remarked that: "The modern mind divides, specializes, thinks in categories. The Greek instinct was the opposite, to see things as an organic whole."[27]

In these studies, and in the history lessons that follow in the subsequent grades, the telling of biographies forms the foundation for dealing with the various cultures. Some obvious examples are Akhnaton and Tutankhamen; Orpheus, Theseus, Odysseus, Socrates, Aristotle, and of course, Alexander. Later comes Hannibal, Julius Caesar, Charlemagne, and others. The children are deeply moved by the struggles of such great figures, and here we see another step taken in fostering a sense for moral judgment and respect for the variety of human endeavor.[28]

## Ways Toward Comprehending the Evolution of Consciousness

Several things assist the teacher in presenting a picture of the development of human consciousness. In his study of Waldorf history methods, Werner Glas found that:

> As each life might be regarded as a symptom of a time, we might call this the symptomatological approach to history teaching. A sequence of such symptomatological lessons would reveal a metamorphosis of consciousness. Metamorphosis is an idea not easily experienced. All too easily the child will appreciate figures of the past in terms of his own age. If we want the child to understand metamorphosis in man, it is good first to introduce him to metamorphosis in nature. For this reason, the natural history period, a botany period, should precede the treatment of ancient civilizations.[29]

Just as teachers can form a link between what the child perceives in the plant world and the changes in cultures, so also can they point out the many things in modern life that stem from ancient times. Some examples are such Greek words as telephone, microscope, photography; in Roman road-building and other engineering, and such Latin words as jurisprudence and justice. The culture and history of Rome moves our study to the sixth grade, where Roman and medieval history are studied.[30] Steiner emphasized that children should gain a feeling for the things that work in evolution out of the depths of the course of time, but that the building of this feeling demands that the teacher postpone the teaching of cause and effect until the twelfth year (sixth grade).[31] Rather, the teacher gives pictures of outstanding personalities and events. As we saw in Chapter III, Steiner always pointed to the hidden streams behind historical events. He maintained that through the imaginative abilities of the teacher, children can be given a feeling for this, and that at the age of twelve, the children can then begin to consider the workings of historical cause and effect. Thus the sixth grade study of Roman and medieval history initiates this consideration.

In history, as in the story curriculum, the teacher can construct for the children a vivid world of images. In this period of development—from the change of teeth to puberty, or from the seventh to about the fourteenth year—the child learns mainly through his feelings. Thus Steiner said, the lessons must be presented artistically, and the child should work artistically. An example of this can be found in the lesson books the children make for each main-lesson subject, and which they illustrate and design. What the teacher presents or tells, they work with in an artistic way.

In the seventh and eighth grades the children move on through time to modern America. The seventh grade undertakes the study of the mighty Renaissance figures, the birth of science, and the great voyages of discovery. The eighth grade studies such topics as the French Revolution; the Industrial Revolution and the impact of technology; the historical effects of nationalism and racism; and differences in life in the nineteenth and twentieth centuries.

It is gratifying to see that a number of teachers are aware that in the United States, we have children from many different cultures. These teachers are striving to reflect in their lessons this diversity of cultures while not losing sight of the goal of preparing the children for life in the Western culture of this nation.

During the fifth through eight grades, many opportunities arise for dealing with the environment—the denudation of the Greek mountains and the resulting erosion; the Nile and its annual flooding; the forests and soils of Europe and America: the environment comes to form a basic part of the historical picture. We should now turn to the study of geography. Before doing so, however, I must characterize the foundation of environmental studies in the curriculum.

### The Ninth Year—A Developmental Turning Point

Franz Carlgren relates the following story:

> Victor was nine. He had always been easy to manage both at home and at school. One day the parents had reason to complain to the class-teacher that Victor had suddenly become extremely obstreperous, refused to take part in the usual Sunday walk, and was most unruly. The music-teacher too, had been having trouble. Victor had always been quite a support in the class, but now he had been blowing his recorder upside down, and doing many other silly things. Even his handwriting had changed to a kind of miniature writing. After four or five months this phase passed. He became quieter again, and the unruliness came to an end.[32]

At this age, children often test the limits of authority, not, as is often thought, to evade it, but rather to ascertain if it is really there. Werner Glas surveyed the work of such psychologists as Jean Piaget, Arnold Gesell, Susan Isaacs, and Dorothy Eichorn; he found a striking unanimity of opinion as to the importance of the psychological changes during the ninth year.[33]

Rudolf Steiner pointed to the ninth year as one of the crucial times in a child's life. He said:

When a child reaches his ninth or tenth year he begins to differentiate himself from his environment. For the first time there is a difference between subject and object; subject is what belongs to oneself, object is what belongs to the other person or thing; and now we can begin to speak o external things as such, whereas before this time we must treat them as though these external objects formed one whole together with the child s own body.[34]

Before this time, the child identifies strongly with his environment, which he experiences as a unity, much like ancient peoples did. As Glas describes it:

Another aspect of these first three years of primary school is that reality has not yet the hard contours of later years.

Dreams and inner impressions slowly give way to the waking consciousness which enables the child to distinguish himself from the world. It is this new wakefulness which produces a sense of separateness in the child of nine . . . . The child becomes aware of the self and the not-self.[35]

He continues:

The nine-year change alters a child's relationship to knowledge. He no longer experiences himself as the unifying factor in the world which now falls into clearer, but more divided, realms of experience. During the tenth year, the child experiences himself as subject and all else as object. Or, to use philosophic terminology for stages of feeling, while previously his apprehension of the world was monistic, he now experiences a dualism.[36]

We have here one of the great issues of philosophy, one which goes to the very root of epistemology—of how we know the world. The child probably will remain in such a dualistic frame of mind—and indeed our culture has struggled with it for several centuries—if steps are not taken to increase his awareness of his relation to the world. Between the years of nine and twelve, the Steiner school curriculum attempts through the teaching of geography, zoology, botany, and mineralogy to give the child a new image of humanity and its relationship to the animal, plant, and mineral kingdoms, so that he can form a realistic picture of the human being in the midst of creation.[37]

If a proper foundation was laid in the will during the kindergarten years, and if a store of living images was built up in the first three grades, then the following three grades can become a healing re-integration of the emerging observer and his world, and lead to an increase in self-knowledge and knowledge of the world.

**Two Aspects of the Third Grade**

There are two areas of the third grade curriculum which are particularly helpful in dealing with the ninth year change. As the child deepens his perception of himself and the world, a sense of loss can arise. The world order no longer seems certain and harmonious, parents and teachers are critically judged and often prove too human or fallible, and as in the above story of Victor, aggressive self-assertion is very common. To provide a series of images for the child, *Old Testament stories* are told.

The Old Testament and Hebrew myths can form a living bridge from imaginations of great cosmic events, to myths that tell of humanity's psychological development, to actual history. Francis Edmunds writes:

> Beginning with the separation of Adam from his God, we follow, through Noah, Moses, the Patriarchs, and so on, the progressive descent of man into earthly existence. Story after story describes how the individual stands in life to realize and serve that which is greater than himself—that which is 'of God' in him. The sense 'of self' in man finds support in a 'sense of origin' and a 'sense of mission'. Experience has proved time and again what strength, comfort and encouragement flows to the nine-year old from this source, how his enthusiasms are engaged, his fears allayed.[38]

In addition to the Old Testament stories, the children study *farming*. This block is a culmination of two years describing the home and its surroundings, but in a special way: in the style of fairy tale and table, where animals and plants talk and act, so the teacher creates and tells stories of everyday life; not as adults see it, but as if the world were *animated* (or "ensouled", in Aristotle's meaning of the word). The environment is spoken of as a living being of which the plants and animals are an integral part. Steiner said that the teacher must be able to transform everything into a fairy tale, a fable, into living substance. He continued:

> Ideally, after very careful preparation, for this kind of teaching demands special preparation, the teacher should be in a position to create out of his own resources conversations between different plants so that the story of the rose and the lily or the conversation between the sun and the moon reach the children directly out of the teacher's individual imagination. . . .
>
> Up to the end of the ninth year everything the child learns about plants, animals and minerals, about the sun and moon, mountains and rivers, should be rendered in this way, for the child is still united with his surroundings.[39]

Steiner underscored the importance of such a presentation for the child's development:

> We must give him the feeling in an artistic way that just as he himself can speak, so everything that surrounds him also speaks.
>
> The more we enable the child thus to flow out into his whole environment, the more vividly we describe plant, animal, and stone, so that in them weaving, articulate spirituality is wafted towards him, the more adequately do we respond to the demands of his innermost being in these

early years. They are years when the feeling life of the soul must flow into the processes of breathing and of the circulation of the blood and into the whole vascular system—indeed into the whole human organism. If we educate in this sense, the child's life of feeling will be called on in a way right for our times, so that the child will develop strongly and naturally in its organism and life of soul.[40]

In the third grade the children meet the farm, and the farm family. Francis Edmunds gives a beautiful characterization from his many years as a teacher:

> The farm is a community of life; it includes the creatures, the plants, the soil itself; it relates to the great cycle of the seasons and leads the gaze from earth to heaven and back again; it comprises so many 'human activities', in the field, in the barn, in the home, at the mill—so many forms of service in which man is seen at his best, as the simple servant and representative of God—the master who serves the good of all; it reveals man as a being who is *more* than nature and yet who stands fully within nature—one who can *order* life for mutual benefit and blessing—a husbandman.(Edmunds' emphasis)[41]

Thus, as the "I" of the children begins to differentiate itself from the environment, the children receive stories and examples of humanity's relationship to God and the world. For example, the third-graders gather stones and build a model farm-house. They construct each room, make mortar to set the stones, and view the house as an extension of humanity and their human and social needs: for working, sleeping, washing, etc.[42] With the foregoing considerations as a foundation, the children are enabled to take up the study of their environment in a new way.

**Geography**

In practice, history and geography cannot be separated completely. Nevertheless, in the Steiner schools, geography forms a separate block of study, and it takes a different perspective than history does. In the history lessons, which begin in a sense with the story curriculum and continue through cultural studies and the biographies of leading figures, the child is led gradually from a world of archetypal images into the experience of various cultures and ages, culminating with the "story" of his own nation. Proceeding from a beginning beyond space and time, the children learn how they came to be citizens of their own time and place. On the other hand, geography lessons take as their point of departure the very place and time in which the children live, beginning with their classroom and their local environment. Like the sciences, the geographical studies are prepared for by the farming and house-building blocks of the third grade, which, as I described, are themselves a certain culmination of the fairy and folk tales and fables.

Werner Glas writes that:

> Just as a stone dropping into a pool makes ever widening eddies, so the teacher, perhaps commencing in the very classroom in which the lesson

> takes place, awakens the spatial awareness of his pupil and helps him to become conscious of ever-widening circles in his surroundings.[43]

For the first time, the child draws a map, a task for which the measurements made during house-building have prepared him. Some examples of mar-making are: a picture of the route the child took to school; his own home and neighborhood; and later, larger areas such as his state, showing rivers and other transportation routes.[44] These maps, as in the case with all work in the Waldorf schools during the first eight grades, emphasize the artistic element. A beautiful collection of such work—for all grades—is found in Franz Carlgren's *Education Toward Freedom.*

In grades four through six, economic geography is taught, including field trips to farms, mills, churches, factories, government buildings - in short, the places where humanity takes the gifts of the earth and works with them. Thus a further step is taken toward helping the children to acquire a feeling of responsibility for the earth on which they live. The fifth grade moves from the local area to a regional approach, studying a geographic whole of which the school's area is a part. As the intellect begins to awaken in grade six, the teacher introduces meteorological and mineralogical concepts, and the children are ready and able to draw maps that represent large areas that reach beyond their immediate experience—this ability to deal with abstraction marks an important point in child development. The sixth grade lessons move from a regional approach to a global one and deal with life in diverse areas such as the equatorial rain forests and the Arctic tundra. In grades seven and eight, the economic perspective gives way to a cultural one. For example, the class may read and discuss accounts of travelers in China, India, Scandinavia, or Africa.[45] At the end of the eighth grade, the children have reached a certain stopping point in their geographic studies. They have begun with their immediate environment and expanded their horizons to the limits of the planet. Many opportunities have emerged to study humankind's relation to the earth and our responsibility for it, from the way people work in the local area, to the economy and culture—the spiritual life—of distant, quite different peoples.

## Zoology and Botany

The study of hunanity's relation to animal, plant, and mineral is undertaken in the fourth, fifth and sixth grades, respectively. Again, we find that fairy tale, folk tale, and especially the fables have formed the foundation for nature studies, the first of which is zoology. A. C. Harwood wrote:

> In an age when men have come to regard consciousness as fortuitous and themselves as an accident on an accidental planet, teachers—at any rate those who know better—will realize the importance of giving a human view of nature as the succeeding stage to the spiritual picture to be found in fairy tales and myths. It is here, above all where Steiner's whole philosophy of life makes a unique contribution to the curriculum. As their first definite study of Nature the children can begin with the most spiritual kingdom—with man himself.[46]

The children begin in the third grade a study of the human form; not physiology per se, but rather

> . . . they should realize what it means for man that his upright position sets his senses free from the service of the body; that his hand is liberated above all special function, being neither webbed, nor hardened into a hoof, nor purely prehensile, nor padded and clawed, but a wielder of manifold tools, a master of the delicate instruments of art, the revealer of the spirit, the bestower of the gifts of love.[47]

Rudolf Steiner remarked that "there is no more beautiful symbol of human freedom than the human arm and hand."

The teaching is always kept in pictorial form. in the fourth grade, the study of zoology begins, As Steiner said "The animal kingdom is experienced as a spread-out human being, as a man spread out fan-wise over the earth." He stated that: "The separate fragments of man, scattered over the earth in the realm of the animals, are in man gathered by the spirit into a total being. Thus we relate man with the animal world, but man is at the same time raised above the animals because he is the bearer of the spirit."[48]

The animals of fable and story are now viewed according to their specialized functions, such as the woodpecker with his powerful beak, the bear's great strength and surprising agility. Each animal's form gives expression to a special ability that is but one of the "many gifts united in the universal form of man." As Harwood described:

> The mouse is all trunk, the little head sunk into alignment with the body, the legs so insignificant that it appears to glide rather than run; the teeth which serve the trunk in the head are always growing, and the mouse must always be gnawing to prevent the stoppage of the mouth by the overgrowth of the teeth. Contrast with the mouse an animal like the horse. How proudly the neck arches, the head and neck becoming like another limb, free and flexible. And how his limbs have grown! Where is his heel? He is always standing high on his toes like a dancer, or like a runner lightly touching the ground with toes and fingers, waiting for the signal to start a race. A bird on the other hand has no real legs at all. When he alights on the branch of a tree you can see that his legs are really only twigs which he has flown away with.[49]

These kinds of descriptions allow the children to use their own imagination and creative talents rather than to imbibe mere facts. There is an extensive literature of animal poetry to augment the artistic presentation of the animal kingdom.

The botany studies of the fifth grade relate the plant to the earth, similarly to the way the animal and the human being were discussed. Here the child's growing acuteness of observation is utilized. The teacher shows the importance of the sun in plant life, and this leads to the great cycles of the seasons and their effects—the first buds of spring, the lush, dreamy green summer, the autumn hues, and the crystalline clarity of winter. The class discusses the different plant parts—the delicate color of blossom and the seed pointed toward the sun, the soft sap-filled leaf and stem, and the hard, earth-covered roots. Sun and earth form a polarity, and the soil with its moisture and fertility nourishes the plant. Different soil structures are examined in a simple way to show how different plants require certain environments.

Steiner pointed to plant study as having quite far-reaching implications:

> If we can give the child this conception of the weaving activity of the earth whose inner vital force brings forth the different forms of the plants, we give him living and not dead ideas. Ideas must develop as a limb of the body. A limb has to develop in earliest youth. If we enclose a hand for instance in an iron glove it could not grow . . . But nothing is more hurtful to the child than definitions and sharply contoured ideas, for these have no quality of growth. Now the human being must grow as his organism grows. The child must be given mobile concepts—concepts whose form is constantly changing as he becomes more mature.

He continued to say:

> Rather must we give the child an idea of what is living in nature. Then its soul will develop in a body which grows as nature herself. We shall not then do, as is often done in education: implant in a body engaged in a process of natural development, elements of a soul-life that are dead and incapable of growth. We shall foster a living, growing physical organism—and this alone serves as a true development.[50]

The third grade farming block can be reprised in a much deeper consideration of the effects of proper farming techniques on soil and plant, and how our greed or benevolence can determine the health of our environment. The discussion of the various plant-communities and their environments forms a natural bridge to geography and history—to the study of human communities.

The botany and zoology blocks are of crucial importance in fostering a view of nature that is life-affirmative. Great emphasis is placed on observation and on an artistic presentation. Perhaps it is well to recall that the great scientific discoveries of the seventeenth century came when poetry still dealt with the same world as that which men contemplated scientifically. Hence it was also the great age of bold and comprehensive scientific views which gave the people of those times a great and glorious new picture of the world. Since that epoch poetry and science have crept apart, poetry concerning itself with subjective experiences, and science with the collection of facts with no comprehensible theory behind them. It is not for education alone that Rudolf Steiner has shown the way to make life whole and sound again by the uniting of the twin experiences of science and art.[51]

These two subjects—zoology and botany—attempt to relate the human being and animal, and plant and earth. They mark an intermediate stage between the more pictorial world of the first school years and the more conceptual, intellectual approach of the sixth, seventh, and eighth grades. And they place a premium on the teacher's creativity. Rudolf Steiner characterized the results of such teaching in this manner:

Botany and knowledge of the plants, taught in the way I have indicated, work on the living world of ideas, and through his good sense place the human being rightly in the world, able to do his work, so that with his ideas he finds his way livingly through life. On the other hand, an equally living conception of his own relationship to the whole animal world specially strengthens the will . . . . The will grows inwardly strong if a man in his own knowledge realizes that by grace of the living spirit, he himself develops as the synthesis of the animal kingdom. This goes into the will, into the forming of the will.[52]

### Mineralogy

Around the age of twelve, a foreshadowing of adolescence appears in the form of the first glimmerings of the intellect, of discernment and the ability to form abstract concepts. Steiner pointed to the physiological changes that accompany the emerging intellect:

> Between the eleventh and twelfth years the skeleton adapts itself to the outer world. . . . If you observe children under eleven years old you will see all their movements still come out of their inner being. If you observe children over twelve years old you will see from the way they step how they are trying to find their balance, how they are inwardly adapting themselves to leverage and balance, to the mechanical nature of the skeletal system.[53]

The study of the mineral kingdom—of inorganic, lifeless matter—comes in the child's twelfth year, at the time when the first presentation of cause and effect in history is given. The discerning reader will notice a progression here. The kindergarten stressed will activity which receives its bodily foundation, according to Steiner, in the metabolic-limb system. The primary school imbues its lessons with the pictorial images that stimulate artistic feeling, which, Steiner held is supported by the rhythmic system. It is thus no accident that the first training of the intellect proceeds through mineralogy,cause and effect in history, and physics (an introduction to acoustics, optics, and heat mechanics). Just as thinking finds its somatic basis in the nerve-sense system, the least organic part of the body, so the first real study of the inorganic world begins as the intellect makes its appearance.

Rudolf Steiner described this important time in the children's life:

> Between the eleventh and twelfth year, and not until then, we may begin to teach about the minerals and stones. The plants as they grow out of the earth are in this way related to stone and mineral. Earlier teaching about the mineral kingdom in any other form than this utterly destroys the child's inner mobility of soul. That which has no relationship with man is mineral in its nature. We should only begin to deal with the mineral kingdom when the child has found his own place in the world when in thought and especially in feeling he has grasped the

> life of the plants and his will has been strengthened by a true conception of the animals the two kingdoms of nature which are nearest to him.[54]

Mineralogy is in effect a part of geography and brings to light in great detail the human relationship to various parts of the world.

## The Study of Humanity and World in the Later Grades

In the seventh and eighth grades, actual physiology is studied. The children examine humanity from the viewpoint of nutrition and health, with obvious opportunities for bringing agricultural connections into the lessons. Various parts of the earth are examined in grade seven, and the earth as a totality in the following year. In these two years the natural history lessons provide the foundation for a study of the conditions in industry, transportation and commerce related to chemistry and physics. There are also a number of ways that ecology and agriculture can be brought into the twentieth century history block in the eighth grade.

## The High School

The high school works with the intellect. These four years are a culmination of the first eight classes, in which the great store of images and observations is called forth out of memory, re-enlivened, and considered in the light of the new powers of thinking. The class teacher gives way to the specialist-teacher, and lessons largely become conversations and discussions.

The natural sciences are "particularly suited to exercising the powers of observation and thinking."[55] Yet the teacher faces somewhat the same challenge that scientists themselves face: to what extent can they present theoretical models as opposed to observable phenomena, models that are a substitute for physical reality, and still help the students to remain free of falling into a belief in the models as reality itself.

Carlgren writes that: "It is a tragic irony that the unheard of scientific advance which more than anything else has led to the erasing of old beliefs, at the same time has brought about a belief in authority which, for its blindness, can be compared to the old forms of religious belief."[56] Mark Twain gave his own inimitable twist to the problem: "There is something fascinating about science. One gets such wholesale returns of conjecture out of such a trifling investment of fact."[57] Yet this challenge means that the high school teachers have an opportunity to lay before their classes, based on the earlier studies, exactly what problems we face today.

In geography, the focus in high school allows an approach that "contributes toward a genuine and scientifically well rounded appreciation of events concerning the entire earth."[58] Geology, supplemented by field trips, shows an ever-changing earth where plants and water-flow transform stone into fertile soil, where mountains arise and erode, where whole continents drift. They study the ocean currents, the patterns of plant life, the atmosphere and the constant bombardment of cosmic substances such as meteors. As the earth "breathes" hydrogen into space, a daily supply of cosmic substances at a rate of 1,000 to 10,000 tons streams inward. So the students can perceive the earth as an organism, a living entity, toward which a feeling of responsibility can arise.[59]

In the eleventh grade, the story of Parsifal is often taught. Parsifal, the "perfect fool," provides many useful and dramatic themes for discussion with the eleventh graders. These include the Holy Grail and the implications, from Steiner's work, of humanity's responsibility for redeeming the earth, and of the interrelationships between human beings and the responsibility for fostering true brother/sisterhood. Those teachers who have taught Parsifal have found that the eleventh graders perceive in a deep and incisive way the relevance for their lives and for the earth of this great story.[60]

### Summary and Conclusion

The Waldorf curriculum is based on an exact understanding of the psychosomatic nature of the human being, given through the work of Rudolf Steiner. The child is helped gradually down to earth from his pre-earthly life, and then assisted to mature into a creative, artistic individual. The language arts and history curriculum enables the child to come to see herself as an individual amidst the forces and events of the past and present. Geography and its relation to the natural sciences allows a widening perspective of the living earth and its inhabitants. Steiner summarized it thusly:

> We must strive to educate in such a way that the intellect, which awakens at puberty, can then find its nourishment in the child's own nature. If during his early school years he has stored up an inner treasury of riches through imitation, through his feeling for authority and from the pictorial character of his teaching, then at puberty these inner riches can be transmuted into intellectual activity. He will now always be faced with the task of *thinking* what before he has *willed* and *felt*. And we must take the greatest care that this intellectual thinking does not appear too early. For a human being can only come to an experience of freedom if his intellectuality awakens within him of itself, not if it has been poured into him by his teachers. But it must not awaken in poverty of soul. If he has nothing within him that he has acquired through imitation and imagery, which can rise up into his thinking out of the depths of the soul, then, when his thinking should develop at puberty he will find nothing within himself to further his own growth, and his thinking can only reach into emptiness.[61]

Steiner maintained that the presentation of a living earth and cosmos could have far-reaching consequences not only !or attitudes toward life, but even for physical well-being, and he based these conclusions on his threefold and fourfold models of psychosomatic processes. He emphasized the "priestly feelings" which the image of the child descending into life from the spiritual world should evoke in the teachers, feelings which will enable them to shoulder the extremely heavy burdens that the Waldorf methods and outlook place on them: "to awaken the child's inborn faculties and guard and foster its unfolding wonder for the world around."[62]

We can see here some parallels between the Waldorf teacher and the biodynamic farmer. Both professions require a great breadth of knowledge in many spheres. Both demand a reverent, even a religious attitude toward life and work, and in both fields, self development and professional development merge. The teacher and the farmer need a knowledge of cosmic and earthly forces and rhythms, and a thorough appreciation of the laws of metamorphosis —of plants, in history, in the child's development, in the farm as an organism. Neither profession pays at all well, especially considering the vast amount of work involved, nor can they expect much understanding from their professional "peers" in orthodox education or agriculture.

Yet there are many consolations and "gifts" that derive from the work. For the teacher, as Steiner described:

> . . . in each individual child a whole world is revealed to him, and not only a human world, but a divine-spiritual world manifested on earth. In other words the teacher perceives as many aspects of the world as he has children in his charge. Through every child he looks into the wide world. His education becomes art. It is imbued with the consciousness that what is done has a direct effect on the evolution of the world. Teaching in the sense meant here leads the teacher, in his task of educating, of developing human beings, to a lofty conception of the world. Such a teacher is one who becomes able to play a leading part in the great questions that face civilization.[63]

The teacher and the gardener are both cultivators, one of the souls of children, the other the steward of the earth. Rudolf Steiner inspired the first Waldorf teachers with these words:

> . . . we are in the same situation as the gardener in relation to his plants. Do you think the gardener knows all the secrets of the plants he tends? No, these plants contain many, many more secrets than the gardener understands, but he can tend them, and perhaps succeed best in caring for those he does not yet know. His knowledge rests on practical experience, he has "green fingers". In the same way it is possible for a teacher who practices an art of education based on reality to stand before children who have genius, even though he himself certainly is no genius. For he knows that he has not to lead his pupils toward some abstract ideal, but that in the child the Divine is working in man, is working right through his physical-bodily nature. If the teacher has this attitude of mind he can actually achieve what has just been said. He achieves it by an outpouring of love which permeates his work as educator. It is his attitude of mind which is so essential.[64]

"Truly we need to find the child in ourselves if we are to *know* children," writes Francis Edmunds (his emphasis). He sees two dangers resulting from the rationalistic-materialistic education of our day: either "a generation of men whose thinking is so far earth-bound that they dismiss all forms of higher perception, including the arts, as unrealistic dreaming"; or the opposite, a retreat from the material world through drugs or mysticism. The hope for the future, Edmunds says

> lies in bringing new powers of imagination into our Western thinking so that human thought can once again testify to a world of higher realities which permeate our everyday living. To serve towards this end is perhaps the highest endeavor of Waldorf education—to lead through education towards the new enlightenment which all the best endeavors of our day, in art, in science, in religion are seeking. This places a great demand upon the teacher to pursue a disciplined way towards a higher grasp of human and natural existence. Unless we as adults can learn to reach up to our children as they descend towards us from worlds before birth, they, as they grow older, will find little or nothing to reach up to in us.[65]

Edmunds describes the "attitude of mind" that Steiner mentioned above in this way:

> Just as an actor, whilst retaining full hold on himself, has by an act of imagination, to one himself with the character he is portraying, so, too, the teacher in his art, whilst retaining the full sense of his adult responsibility, has to one himself by an act of imagination with his children, has to become 'child' with them at whatever level he may meet them so that the content of his lessons may interpret life according to their need. (Edmunds' emphasis)[66]

He concludes that the teacher needs to practice a true art of education: "He needs to be of the order of the magi, an interpreter of wisdom, a magician who can conjure up reality in a thousand forms, a servant of truth drawn from a higher source of reason."66

From the work of Rudolf Steiner, and those who have followed in it, it appears that both the farmer and the teacher have to become artists, to make their professions into an art, through which they attain to a higher wisdom in their work. Thus the farmer enlivens his soil and the teacher guides his children.

**Footnotes for Chapter Seven**

1. Rudolf Steiner, *The Education of the Child in the Light of Anthroposophy*, 2nd ed. (London: 1975) 40.
2. E.E. Pfeiffer, *Soil Fertility, Renewal, and Preservation* (London: Faber and Faber, 1949) 29.
3. Steiner, *Education of the Child*, 3.
4. The quote is from Stewart C. Easton, *Man and World in the Light of Anthroposophy* (New York:1975) 384; Herbert Hahn, "Birth of the Waldorf School from the Threefold Social Movement," *Golden Blade* (1958) 58, 68-69. The three education courses are: *The Study of Man* (London: 1966); *Practical Course for Teachers* (London: 1937) and *Discussions with Teachers* (London: 1967). For the life and work of Emil Molt, see Christine Murphy, ed. *Emil Molt and the Beginnings of the Waldorf School Movement* (Edinburgh: Floris Books, 1991).
5. Easton, Ibid., 385. The more important of Steiner's work on education, in addition to the three basic courses in note 4 are: *A Modern Art of Education*, 3rd ed. rev. (London: 1972); *Human*

*Values in Education* (London: 1971); *The Roots of Education* (London: 1968); *The Spiritual Ground of Education* (London: 1947); *The Kingdom of Childhood* (London: 1964); *Education as a Social Problem* (New York: 1969); *Soul Economy and Waldorf Education* (New York: 1986); *The Child's Changing Consciousness and Waldorf Education* (New York: 1988); and *The Four Temperaments* (New York: 1968).

6. Steiner, *Education of the Child*, 6.

7. Perhaps the most basic presentation of the fourfold aspect of humanity is found in Steiner's *Theosophy* (New York: 1971). For the term "body", see 16-17.

8. Steiner, *Education of the Child*, 9.

9. Ibid., 12,

10. Ibid., 12-13.

11. Ibid., 14-15. The observant reader will recall the experience of the "I am" of Moses, and of Christ's "I am" statements in my chapter four.

12. See my chapter two, and A.C. Harwood, "Life Forces and Death Forces," in his *The Way of A Child*, 4th ed. rev. (London: 1967) 69-81. An excellent discussion of the medical applications arising out of these models of the human organism will be found in Easton, *Man and World*, 471-478.

13. Steiner, *Education of the Child*, 23.

14. Easton, *Man and World*, 392; Werner Glas, *Speech Education in the Primary Grades of Waldorf Schools* (Wilmington, Del.: Sunbridge College Press, 1974) 7-8; *John Gardner, The Waldorf Approach to Education: Questions and Answers* (Sacramento: Sacramento Waldorf School, n.d.) 12.

15. Glas, Ibid., 11.

16. Francis Edmunds, *Rudolf Steiner's Gift to Education: The Waldorf Schools*, 3rd. ed. rev. (London: 1975) 137.

17. Gardner, *The Waldorf Approach*, 2.

18. Susanne Langer, *Philosophical Sketches* (New York: Mentor Books, 1964) 80, Quoted in Glas, *Speech Education*, 23.

19. Ibid.

20. Easton, *Man and World*, 398-399.

21. Ibid., 385; Frans Carlgren, *Education Towards Freedom: Rudolf Steiner Education—A Study of the Work of Waldorf Schools Throughout the World* (East Grinstead, Sussex: Lanthorn Press, 1976) 71.

22. Carlgren, Ibid., 97.

23. Glas, *The Waldorf School Approach to History* (Detroit: Waldorf Institute, 1963) 8.

24. Perhaps the best single account of the story curriculum is in Werner Glas, *Speech Education*, 41-67.

25. Ibid., 64.

26. Ibid.; Glas, *History*, 10-11.

27. H.D.F Kitto, *The Greeks* (Baltimore: Penguin Books, 1967) 169, quoted in Glas, *Speech Education*, 65.

28. Glas, *History*, 21

29. Ibid.

30. Ibid., 21-22.

31. Steiner, *The Kingdom of Childhood*, 65.

32. Carlgren, *Education Towards Freedom*, 98.

33. Ibid.; Glas, *Speech Education*, 24-26.

34. Steiner, *The Kingdom of Childhood*, 49.

35. Glas, *Speech Education*, 24.

36. Ibid., 25.

37. Ibid.

38. Edmunds, *Rudolf Steiner's Gift to Education*, 61.

39. Steiner quoted in E.A. Karl Stockmeyer, *Rudolf Steiner's Curriculum for Waldorf Schools* (Stuttgart: Verlag Freies Geistesleben, 1973) 155-154.

40. Steiner, *A Modern Art of Education*, 138-139.

41. Edmunds, *Rudolf Steiner's Gift to Education*, 62.

42. Ibid., 63.

43. Glas, *History*, 5-6.
44. Carlgren, *Education Towards Freedom*, 104.
45. Ibid., 104, 118-120; Glas, *History*, 6-7.
46. Harwood, *The Way of A Child*, 60.
47. Ibid.
48. Steiner quoted in Carlgren, *Education Towards Freedom*, 108; Steiner, *A Modern Art of Education*, 148.
49. Harwood, *The Way of A Child*, 60-61.
50. Carlgren, *Education Towards Freedom*, 112. The two quotations by Steiner are from *A Modern Art of Education*, 144-145.
51. Harwood, *The Way of A Child*, 61, discussing Buckle's *History of Civilizationf in England.*
52. Steiner, *A Modern Art of Education*, 150. See the important article by Günther Hauk, "Gardening as a Subjet in the Waldorf School," *Biodynamics*, 169(Winter 1988-1989) 21-28.
53. Steiner, *The Spiritual Ground of' Education*, 76, quoted in Glas, *Speech Education*, 26.
54. Steiner, *A Modern Art of Education*, 169.
55. Carlgren, *Education Towards Freedom*, 145.
56. Ibid., 160, 145.
57. Mark Twain, *Life on the Mississippi* (New York: Harper, 1913) 156.
58. Carlgren, *Education Towards Freedom,* 161.
59. Ibid., 161-165.
60. See Franz Winkler, *Man: The Bridge Between Two Worlds* (New York: 1960), a study of the Holy Grail and Parsifal story; and René Querido, *The Holy Grail: A Modern Path of Initiation* (Fair Oaks: Steiner College Publications, 1992).
60. Steiner, *The Roots of Education*, 84-85.
61. Eileen Hutchins, "The Normal Child," in John Davy, ed. *Work Arising From the Life of Rudolf Steiner* (London: 1975) 81.
62. Steiner, *Human Values in Education*, 26.
63. Ibid., 27.
64. Edmunds, *Rudolf Steiner's Gift to Education*, 43-44.
65. Ibid.
66. Ibid.

## CHAPTER EIGHT

### Waldorf Schools and Biodynamics

*Although it may seem absurd, it must be stated that a person who has not learned to distinguish an ear of rye from an ear of wheat is no complete human being. It can even be said that a person who has learned to distinguish between rye and wheat without having observed them growing in the fields, has not attained the ideal. As teachers we should avoid going on botanical expeditions to collect specimens to be shown in the classroom. The children should be taken out and whenever possible, be brought to understand the plant world in its actual connection with the earth, with the rays of the sun, with life itself.*

*We must try to realize what it means for the evolution of humanity that for a long time past large numbers of people have been drawn into the towns, with the result that generation after generation o young people in the great towns has grown up in such a way that they can no longer distinguish wheat from rye.*[1]

*Rudolf Steiner*

A number of Waldorf schools employ gardening as a part of the curriculum, particularly from the sixth grade onwards. If a good garden has been established on the schools grounds, it can have several uses for the different grades.

In the early years, simple observation of garden activities draws the children into the mysteries at work as plants grow, bear fruit, and wither, as the soil is cultivated, and as the gardeners take leaves and old plants to make compost to nourish the next crops. Some schools begin to involve the children in the garden at this time in tasks suited to their abilities. Others begin with the first gardening block in the sixth grade. Some schools follow Steiner's original plan in which gardening is taught from grade sixth through grade ten for two lessons per week.

Steiner once spoke about the value of garden work for children:

> And if you could even make little ploughs and let the children cultivate the school garden, ii they could be allowed to cut with little sickles, or now with little scythes, this would establish a good contact with lime. Far more important than the skill is the psychic intimacy of the child's lie with the life of the world. For the actual fact is: a child who has cut grass with a sickle, mown grass with a scythe, drawn a furrow with a little plough, will be a different person from a child who has not done these things. The soul undergoes a change from doing things. Abstract teaching of manual skill is really no substitute.[2]

Steiner placed great emphasis on such practical work, which he thought should follow the more artistic work. As the child enters adolescence, work in gardening and handicrafts such as woodcarving should help him really to enter the practical world. As he said:

> During the ages from 15 to 20 everything to do with agriculture,trade, industry, commerce will have to be learnt . . . . During these years all those

subjects will be introduced which I would call world affairs, historical and geographical studies, everything concerned with nature knowledge—but all this in relation to the human being, so that man will learn to know man from his knowledge of the world as a whole.[3]

It is significant that Steiner recommended that garden work as part of the curriculum begin during the twelfth year—also the time when mineralogy and cause and effect in history are taught, and which I described in the preceding chapter as a significant turning point. Franz Carlgren writes that "The independent life of feeling awakens and changes the relationship to one's own body, to the environment, and to ideas and ideologies. It reflects itself in one's ability to love and be interested in the world, as well as in the ability to reason and to judge."[4] As the intellect emerges, gardening can provide a strong balance or the conceptual work. It is part of the approach of the Waldorf high school in preparing the children for life in the world.

Rudolf Steiner decried the act that increasingly fewer people knew first-hand the world of machines and processes that produce much of their daily surroundings. As they enter adolescence, the children should be introduced to the environment in which men work, but which few people really understand, The economic geography lessons supply a picture of man through a spatial orientation, while the "lessons preparing for life and technology supply social concepts to the picture of the human being. These two subjects give an important complement to the gardening studies. The technological preparation for the lessons are given four afternoons per week for a nine-week block in the tenth grade, and for a three-week block in the eleventh and twelfth grades.[5] Surely today people are even more removed from the functions of the world than they were fifty years ago. The lessons try to give high school students an immediate experience of the world they may be entering soon if they do not go to college, and perhaps even more importantly, they give college-bound students a taste of the "real world".

## The Gardening Curriculum

A typical Waldorf gardening curriculum might approximate the following scheme:

*Grades six through eight:* The children cultivate the soil and tend and harvest vegetables and flowers. During these three years, the repetition affords an experience o the seasons and of crop rotations. The dormant winter period of bad weather can allow time to discuss the spring, summer, and all work. Other indoor work can be done.

*Grade nine:* The students are introduced to more complicated cultivation and plant-tending, and learn to make compost heaps. Fruit-tree and soft fruit work can begin. During the winter months the connections between plant growth and sub-soil, weather and astronomical conditions are studied, as well as the origin of the most important cultivated plants, the necessary conditions for their cultivation and the different methods of reproduction.[6]

Here the gardening teacher can explore a bit of the history of agriculture, and the difference in origin and cultivated plants, which can lead into medicinal herbs.

Plant reproduction studies involve propagation from seed, by striking, grafting, and budding.

*Grade ten:* The previous year's studies and work are continued, plus learning to prune trees and shrubs. The previous winter's study of propagation now can lead into the mystery of tree grafting, of learning to do it. As a complement to the handicraft class, garden tools can be repaired over the winter, and the students can build garden paths and walls. Lessons on proper manuring are a good introduction to modern agricultural problems and to livestock-raising.[7]

The emphasis is on learning to see the farm as an organic whole, a living organism, understanding the art of manuring as in intrinsic part of nutrition, and perhaps even studying the biodynamic preparations as a complement of the organic chemistry block.

## Gardening in Waldorf Schools

Several American Waldorf schools use gardening as part of the curriculum.

The Kimberton Waldorf School opened in 1941, and now comprises kindergarten through high school The school was given the four hundred acre farm in which its grounds sit, and farm work is being incorporated in several aspects of school life. At the present time, the kindergarten and first and second grades observe farm work and sometimes pick a few apples, etc. The third grade farming block involves the children actually living at the school for a week in the spring, when they plant squash, pumpkins and other crops that they can harvest in the fall as fourth graders. The fourth grade zoology block might involve washing the cows and other animal chores, and there are obvious benefits for the botany block in the fifth grade. The farm work expands as the children become more capable. The high school students can find many opportunities for farm work and gardening.[8]

The High Mowing School in Wilton, New Hampshire, founded in 1942, has a small farm connected to it. There is a large garden on the grounds in which the children can work. The school comprises grades eight through twelve. Fortunately, there are several biodynamic initiatives nearby for the children to visit.

The Hawthorne Valley Steiner School at Harlemville, New York is located across the road from a working dairy farm of over four hundred acres (see the description below in "Apprenticeships and Biodynamic Training"). The school now extends from kindergarten through grade twelve. In addition to a small garden for the children to work, the school maintains a program for visiting students directed by Nancy Dill, so that Waldorf students from city schools such as the Rudolf Steiner School in Manhattan or the Waldorf School on Long Island can live and work on the farm for two or three weeks.

The Sacramento Waldorf School in Fair Oaks, California, possesses a fine biodynamic garden. Gardening is a part of the curriculum beginning in the sixth grade. The Highland Hall School in Northridge, California also has a garden. Both these schools comprise kindergarten through high school. It is in urban areas like these that a garden can especially open new worlds for children. Many schools take the third grade class to spend some time on a biodynamic farm as part of the farming block, described in a previous chapter, or in later grades.

A newer venture is the Covelo Farm School in Covelo, California, located about three hours north of San Francisco and managed by Steve and Gloria Decater. They have developed a multi-faceted operation with both educational and agricultural purposes.

> The Decater farm has many functions which include community-supported agriculture, farm trips for school children, training for adults in the use of draft horses and commercial vegetable production, yet it gives the impression of more unity than any farm I've visited in the past few years. This is no doubt in part due to its persistently human scale. . . [It fills] a gap between the homestead and the larger farm, so that even a city dweller can grasp what's going on. The farm has retained that necessary balance of animals and plant life that used to provide a subsistence living for a large family.[9]

Children from a large number of Waldorf schools visit this unique farm, and apprentices have come from the U.S. and other nations. There are few places where one can gain such an immediate appreciation for where farming has come from, and where human beings live in such harmony with soil, plants, and animals. One parent wrote:

> The children, I think, will remember lots of experiences from the farm—how a donkey sounds in the middle of the night and what a bright yellow the butter is at that magical moment when it separates from the whey. But maybe they will also remember, even if the never pull it into words or even consciousness, that they felt a certain at-homeness and acceptance of the natural and manmade—*that things made sense at the farm.*[10] (My emphasis)

There could be no better statement of the rationale for the dedicated work of the Decaters or Nancy Dill at Harlemville, or Ruth Zinniker at her farm in Wisconsin where children visit from the Chicago and Milwaukee Waldorf schools, in providing children with these experiences of the inherent order in the proper relationship between the kingdoms of nature—*right living*, in the Buddhist sense. It takes a special kind of person to be able to do farm work and relate well to groups of children. One visitor said "The Covelo Farm School is a wonderful place and would be worth of imitation, if one could find two people with Gloria's encompassing warmth and Steven's skill and patience."

**Curative Education: The Camphill Movement**

Stewart C. Easton writes:

> Although accurate statistics are not available for earlier centuries it appears to be an unquestionable fact that children who are in some way abnormal, either physically or mentally or both, are being born today not

> only in increasing numbers but as an increasingly large percentage of the population, at least of the Western world.[11]

Easton says that the anthroposophical work with retarded or handicapped children in the past fifty years

> . . . has probably received most recognition from non-anthroposophists, especially in the English speaking countries. Even those who may remain totally skeptical about Steiner's spiritual knowledge are often willing to admit that the curative education developed by these anthroposophists appear to do more for the children than is done by the practitioners of any other form of therapy. Hence it is quite natural that these institutions should receive more state support than is available for Steiner education, since most governments are only too happy to turn *this* work over to private enterprise. (Easton's emphasis)[12]

Lack of space precludes all but the most abbreviated treatment of this area of Steiner's work. It is perhaps easy to see that the medical and nutritive aspects are of utmost importance in caring or these children. For that reason most curative homes and communities make use of their own, or nearby, biodynamic gardens and farms; of the medical and diagnostic innovations begun by Steiner; and of the special massage, color therapy, and curative eurythmy treatments developed by Steiner, his coworkers and later practitioners. The goal is, as in Waldorf education, to educate the child in accordance with his potentialities.[13]

The homes and villages accentuate a "family" atmosphere to form a substitute family around the children. With the severely handicapped and retarded children, the coworkers, the nurses and curative teachers are on call day and night, even when not actually working. Stewart Easton says:

> So it is impossible for them really to have a "private" life of their own except insofar as a married couple can quite easily work in a Home and frequently do. Their work thus becomes their life, and their recreation consists in whatever the Coworkers do together, and with the children. It is entirely natural that the plays and festivals should become the main "recreation"—in the true sense of the word, re-creating themselves.[14]

The Camphill Movement, since its founding in Scotland in 1939 by Dr. Karl König, has rapidly expanded to many parts of the world. Camphill Homes, Special Schools, and Villages are curative institutions run as intentional communities along lines indicated by Steiner and developed by Dr. König and his colleagues. No one receives a salary, and each member's needs are met by the community. This allows the coworkers to devote all their time and energy to the children, free of personal financial burdens. The community as a whole deals with economic matters.[15]

The Camphill Villages care for adult handicapped and retarded persons. It is realized that these people probably never can become "normal" members of society. Therefore the Villages attempt to provide a family and social life through the community life that is as normal and fruitful as possible. The Villagers, as these people are known, receive training in a handicraft, or some other useful work in the Village, and they have full employment for as long as they wish it. The Villages are estates with various workshops, gardens, and residences in which the Villagers and coworkers live together. The quality of goods the Villagers make is quite high and brings in income to the Village. The Camphill Villages strive for self-sufficiency, and the fact that their labor contributes to this goal becomes a great source of morale to the handicapped, since they can feel that they are making a real contribution to the community.[16]

In the United States, there are four Camphill establishments. At Copake, New York, a Village of over 600 acres comprises a biodynamic farm and garden and several handicraft shops. The Villagers take part in farm and garden work alongside the coworkers, as their abilities will permit.

The Kimberton Hills Village near Kimberton, Pennsylvania does not emphasize craft work, but rather has sought to become a self-sufficient community based on the dairy and cereal farming and the large gardens, both of course biodynamic. It is the first Village to be devoted solely to biodynamic agriculture. The Villagers work in the garden, on the farm, or in the residences. The community is located on an old farming estate that comprises approximately 350 acres. The houses range from several converted farm buildings, old Pennsylvania Dutch farmhouses, new dwellings, and a large manor house. The community has a farm store, a bakery, makes cheese, and maintains a coffee shop for visitors and residents.

About a twenty minute drive from Kimberton Hills one finds the Camphill Special School at Beaver Run. It includes sixty acres of woodland, fields and meadows, gardens, and lovely family-style residences situated on a hillside. The school takes mentally, emotionally, and physically handicapped children from the first grade through high school. Bernard Wolf writes:

> The work on the land during the school year (September-June) is integrated with the school program, and the two draw strength from each other. Working with those who overlook the land work, the Land Panel, are a group of boys of high school age. These children, of varying abilities, share in the service to the land for a few hours each day, and the demands of the land provide work and training for them. Their work program includes composting, mulching, hay and leaf raking, sowing and planting, weeding, hand cultivating, carrying rocks and branches, loading and unloading wheelbarrows and trailers. In the summer, when the children are gone, other members of the community join the Land Panel in the work.[17]

Near the school at Beaver Run is Soltane, the newest Camphill community in the United States, which serves those children who are too old for the school but too young to begin life in a Village. Soltane has a garden, large berry plantings, and fruit trees, in addition to craft work.

An added advantage of community life is that the farmer and gardener can concentrate solely on the demands of the land, on nurturing, healing, and improving soil and plant, without being subject to the whims of the marketplace. Hartmut von Jeetze says, "Once again the farmer is assured of his true position, that of a mediator between a community of men on the one hand and divine forces working in the organism of the land on the other."[18]

He adds:

> Another important aspect of the land is its therapeutic value. Our approach has made it possible for many persons— people who elsewhere would be social outcasts in a world of competitive "profitability"—to find true fulfillment in the social organism of Camphill. In the centers of the Camphill Movement, which integrates handicapped people into creative community life, many mentally retarded persons have been able to find a place meaningful for them, as well as for the social organism of which they are a part, only through being allowed to take their place in the work on the land.
>
> Quite apart from economic considerations, their day-by-day involvement in nature's seasonal processes of growth, dying and rebirth has a therapeutic value which could not be replaced by any other means. Not to avail oneself of this opportunity would be unthinkable in the Camphill approach to man and nature. The social and therapeutic value of work and life with the land is unquestionably reestablished in the striving of the Camphill centers throughout the world.[19]

In Sauk Centre, Minnesota, Camphill Minnesota is a Village which operates a dairy farm of over two hundred acres and gardens.

## Training in Biodynamics

Learning and teaching biodynamics can be approached from two directions: practical training and academic, or classroom, work. Of course it is possible to combine these two approaches, and this has been done in several instances; yet certain problems are inherent in such a combination. Farmers and gardeners often possess a wealth of scientific knowledge of soils, plants, and animals, but their main focus usually is toward the practical aspects and is centered on their particular area, its soil and climate. The farmer's long and constant work-hours may well preclude any organized teaching, no matter how willing or capable he may be at it. Agricultural scientists, on the other hand, usually feel more at home in the classroom than on a tractor or digging a bed of leeks. The two branches of training are complementary, however, since the biodynamic approach to agriculture demands a very wide spectrum of knowledge.[20] In my essay on biodynamics (Chapter Six), I pointed out Steiner's emphasis on the symbiotic relationship he thought should exist between the farmer and the scientist. Each must come to have both some of the old "peasant wisdom" and also the artistic element required by Goethean science.

It is possible, says Carsten J. Pank of Ceres Farm in New York, to learn much of the practical knowledge needed for biodynamics on conventional farms and gardens:

> The majority of successful biodynamic farmers throughout the world

> started and became experienced after they were thoroughly acquainted with conventional farming methods. Aside from mixing and applying the BD preparations, manure handling, and composting, the difference between conventional and biodynamic farming is in the area of management decisions which derive from new spiritual aspects and concepts. In regard to practical work skills and experience, the difference is relatively minor; to milk cows, to adjust a grain drill, to set a plow, etc. —all this can certainly be learned on a conventional farm.[21]

Pank continues:

> I can see no valid reason why, on this continent, the present shortage of experienced biodynamic farmers who have the time and personal interest to teach BD apprentices should prevent anybody from at least getting started in his own farm training. In fact, I personally believe it to be a good idea, and at times practically helpful, if the future BD farmer has at least a basic understanding of conventional methods.[22]

Historically, many U.S. biodynamic farmers have not learned conventional farming first, but have attended training courses and conferences sponsored by the Biodynamic Association, and have been visited by established biodynamic farmers to gain an idea of what biodynamic practices and principles involve. As I indicated above, the difference between conventional and biodynamic farming is to a large extent a matter of farm management decisions based on new concepts and attitudes. These often are acquired through actual practical work, and through reading, lectures, and discussions. Fortunately, there are now two training courses in the U.S., and a good number of farms on which one may apprentice.

In the remainder of this chapter, I will discuss various biodynamic farms and gardens that accept apprentices, the training centers where one can receive instruction in the biodynamic method, and conferences where lectures on biodynamics are presented.[23]

## Apprenticeships and Biodynamic training courses

Christoph Meier manages a 450 acre dairy farm in Harlemville, New York community which is part of the Rudolf Steiner Educational and Farming Association. This includes the Hawthorne Valley Rudolf Steiner School (K-12), the farm, and the visiting students program, which hosts classes from other Waldorf schools. They come for up to a week to experience farm work. There is a residence for them and a dining hall that serves food raised on the farm. The cultural life at Harlemville is enhanced by the school and its teachers, the close proximity to the Camphill Village at Copake, New York, and the lively anthroposophical life of the area.

Christoph was trained in his native Switzerland and in Holland, and he has been very successful in building up the farm along biodynamic lines since 1977. The farm processes its products into cheese, yogurt, and bread, and it also provides vegetables for the busy farm store. In addition to managing a dealership on the farm property for Belarus tractors, Christoph Meier has begun to distribute organic and BD produce in the northeast U.S. For many years Christoph has taken apprentices, teaching them all aspects of biodynamics. Recently a six-week intensive winter course began there for apprentices from Harlemville and other BD farms. It is hoped that this can grow in the coming years.

The Zinniker Farm near Elkhorn, Wisconsin takes apprentices, usually one per year. During the past fifteen years, many young people, often from Europe, have apprenticed with the Zinnikers. They have learned a great deal from Ruth and Dick Zinniker, and have been able to participate in the conferences, Christian Community services, and anthroposophical study groups that are part of the cultural activities there. Children from the Chicago Waldorf School come regularly to spend a week on the farm. Two other BD farms are nearby—Nokomis Farm and the Krusenbaum Farm, owned by the Zinnikers' daughter, Susan, and her husband, Altfrid Krusenbaum. It is a mixed dairy farm. A new venture is the Michael Fields Agricultural Institute, which engages in research and farm advising for biodynamic and organic farmers.

Kimberton Hills Camphill Village near Kimberton, Pennsylvania has provided one, two, and three year trainings since 1978. Apprentices live as members of the community, sharing in all aspects of community life; this provides an insight into the unique arrangements for aiding the economic problems of farmers that have arisen in the Camphill movement. The growing season affords excellent opportunities for practical gardening or farming, and from October through April the apprentices participate in study groups and a variety of lectures concerning biodynamic agriculture, along with the practical work. Apprentices work and live with the handicapped adults and take part in the spiritual and cultural events characteristic of these villages, which include celebration of the major festivals of the year and a variety of lectures and artistic activities. I participated in this course, and I learned many valuable things, as much from the community life with the handicapped people as from the gardening and baking I did, as much about myself as about nature.

Hugh Courtney, who has made biodynamic preparations for the Biodynamic Farming Association for many years, often has a place for an apprentice who wishes to learn to make these preparations.

It can be beneficial to precede either an apprenticeship or a training course with one of the "foundation year" programs of anthroposophical studies available at several institutions. These are multidisciplinary programs that balance conceptual studies, several artistic disciplines, and craft work. In this manner the three areas of soul activity—thinking, feeling, and willing—are engaged and integrated in a meaningful way. The conceptual courses examine the development of human consciousness as it comes to expression in art, literature, philosophy, and social systems. Other courses concern the history of scientific thought, the human being's relation to the kingdoms of nature; fundamentals of psychology related to anatomy and physiology. Also studied are the philosophical basis of Steiner's work, the forces active in social life today, and the evolution of religions. These programs outline Rudolf Steiner's contributions to a variety of cultural endeavors, reviewed in the light of developments in the past decades. Many people do not have the time or resources to make these programs part of their agricultural training;

the advantage is gaining a good foundation in anthroposophical knowledge and meditational practices.

The two oldest and most developed programs are given at the Waldorf Institute in Spring Valley, New York, and at Emerson College, Sussex, England. At Rudolf Steiner College in Fair Oaks, California another such course has grown rapidly and affords an opportunity to West coast residents. There, Harald Hoven directs a garden which provides produce to the College community and medicinal herbs for the nearby Raphael Clinic and Pharmacy. Addresses for these institutions, farms, and other schools mentioned in the chapter are given in the appendix.

The programs mentioned above provide an insight into Steiner's work and the broad interdisciplinary background assists in understanding ten biodynamic approach to nature. Thus the benefits of an apprenticeship or a training course in biodynamics are greatly enhanced.

Emerson College offers a course in biodynamic farming and gardening. The students follow a on year program that includes studies, acquiring specific skills and project work; students are taught the principles of biodynamic and ecologically sound agriculture. They are introduced to the concepts and practice of biodynamic plant and animal husbandry and planning predominately self-reliant farm systems. The understanding and method of production of nutritional quality in food and foodstuff are part of the program. Marketing and the establishment of certification programs are included. Independent judgment in the application of biodynamic and biological methods and how to use resources are aims of the training. The course is designed to achieve, through lectures, practical demonstrations, and work experience including individual projects and studies, an understanding of biodynamics. It is meant to complement practical and theoretical training in agriculture and horticulture.

At this time (1991), the program is being re-designed and interested readers should write the College directly. Many one to three week short courses have been started on specific subjects such as: forestry, climatology and ecological zones, geology, formative factors in the landscape, gardening, animal health, and biodynamic seed production. The Rural Development Program has for many years worked in conjunction with CREAR (Centro Regional de Estudios de Alternativos Rurales) in the Dominican Republic to train students for work in developing nations.This program has links with programs in Asia, Africa, South America, and other places.

## Conferences and Lectures

Several conferences which .feature lecturers on various aspects of biodynamic agriculture are held in the United States and Canada each year. The Biodynamic Farming and Gardening Association holds its annual three-day conference in late summer or early autumn. Each year the conference is based on a theme around which a number of lectures and discussions are built. For example, the central topic for 1990 was "Calcium and Water". Lectures from the conference are often published in *Biodynamics*, the quarterly magazine of the Association. Lecturers and workshop leaders include prominent agricultural scientists, farmers, and gardeners.

The location of this annual conference moves each year so that in the course of a few years it covers the country. Regular regional conferences take place in California, Wisconsin, the northeast, Ontario, Quebec, and British Columbia. Regional groups in many other areas organize one day workshops. The Biodynamic Newsletter contains notices of regional and national conferences.

## Training in Other Nations

There exist in other countries several hundred farms that employ biodynamic methods, and also many market gardens. Accordingly, apprenticeships are more plentiful than in the United States. Master farmer and gardener trainings are well established in several European countries.These are state certified programs that also include biodynamic methods. At the Warmonderhof in Holland, there is a fully accredited training program in biodynamics. New Zealand now has a one year program, as does Brazil; these are open to students from other nations.[24]

### Footnotes for Chapter Eight

1. Rudolf Steiner, quoted in E.A. Karl Stockmeyer, *Rudolf Steiner's Curriculum for Waldorf School* (Stuttgart: Verlag Freies Geistesleben, 1973) 231.
2. Ibid., 322.
3. Ibid., 235,
4. Frans Carlgren, *Education Towards Freedom* (East Grinstead, Sussex: Lanthorn Press, 1972) 132.
5. Stockmeyer, *Rudolf Steiner's Curriculum*, 235-237.
6. Ibid., 233-234. See the important article by Günther Hauk, "Gardening as a Subject in the Waldorf School," *Biodynamics*, 169 (Winter 1988-1989) 21-28.
7. Ibid., 234.
8. The Kimberton area is home to the large farm at Kimberton Hills Camphill Village, the community-supported market garden directed by Kerry and Barbara Sullivan, the garden at Camphill Special School, and the offices of the Biodynamic Farming and Gardening Association..
9. Joel Morrow, "The Covelo Farm School," *Biodynamics*, 169(Winter 1988-89) 4-5.
10. Mary Jane DiPiero, "Whoever Thought I'd Plow Behind a Horse?" *Biodynamics*, 169(Winter 1988-89) 19.
11. Stewart C. Easton, *Man and World in the Light of Anthroposophy* (New York: 1975) 432.
12. Ibid.
13. Ibid., 423.
14. Ibid., 428.
15. Ibid., 412, 427-428.
16. Ibid., 429-430
17. Bernard Wolf, "Agricultural Work at Camphill Special Schools," *Biodynamics,* 108(Fall, 1973) 12. See also Cornelius Pietzner, "Camphill: Celebrating 25 Years in America"; Andrew Hoy, "Anniversary Reflections"; Janet McGavin, "Beginning of the Camphill Work in America"; and Christof-Andreas Lindenberg, "Thoughts on Curative Education and Social Therapy," all in *Journal for Anthroposophy*, 44(Winter 1986).
18. Hartmut von Jeetze, "Agriculture and the Camphill Movement," *Biodynamics*, 114(Spring, 1975) 6.
19. Ibid.
20. The Chambers Family, "The Horns of A Dilemma," *Biodynamics*, 108(Fall, 1973) 2-4.

21. Carsten J. Pank, "A Letter to Those Who look For Employment on a Biodynamic Farm," *Biodynamics*, 118(Spring, 1976) 35.
22. Ibid., 36.
23. The entire issue of *Biodynamics*, 160 (Fall 1986) is devoted to training programs.
24. See Michael Wildfeuer, "A Course for Young Farmers in Dornach," *Biodynamics*, 164(Fall 1987) 37-43; see footnote 23, above.

# CHAPTER NINE: CONCLUSION

## A Brief Summary

*Thus in those times a divine truth, a moral goodness and a sense-perceptible beauty existed in the Mystery Centers, as a unity comprised of religion, art and science. It was only later that this unity split up and became science, religion and art, each existing by and for itself. In our time this separation has reached its culminating point. Things which are essentially united have in the course of cultural development become divided. The nature of man however, is such that for him it is a necessity to experience the three in their "oneness" and not regard them as separate. He can only experience in unity religious science, scientific religion and artistic reality, otherwise he is inwardly torn asunder. For this reason wherever this division, this differentiation, has reached its highest pitch it has become imperative to find once more the connection between these three spheres.*[1]

*Rudolf Steiner*

I began this study with a brief inquiry into the all-encompassing nature of agriculture and its relation to the scientific world-conception. One aspect that emerged from the first chapter is that a one-sided view of this scientific world-conception, riding the wave of the Age of Reason, has resulted in a new belief in authority. As the great physicist and physiologist von Helmholtz warned:

> I beg of you not to forget that even materialism is a metaphysical hypothesis, a hypothesis that proved very fruitful indeed in the field of natural science but nevertheless under any circumstances is only a hypothesis, and if one forgets this, it turns into a dogma that impedes the progress of science and leads to passionate intolerance as do other dogmas.[2]

The reductionism of modern science, which seeks to divide the world into increasingly small fragments in order to isolate their chemical and atomic constituents, runs counter to a growing need of our day. This is the need for a progressively wider, more inclusive view, seen most dramatically in ecology and other earth sciences.

The biosphere with its very complex interweaving of ecosystems requires, it seems, a world-view that can reach out even into the vastness of the cosmos. Such a world-view in turn demands a strengthened and enlivened thinking, new abilities with which to apprehend and comprehend phenomena.

I then presented the natural scientific work of J. W. von Goethe, with emphasis on his methods of observation and his concept of the evolution of human consciousness as a perspective for dealing with the history of ideas. The hypothesis that human faculties have undergone a process of evolution through the ages, and that perception and thinking can be trained and enhanced through self-development, led to an examination of the life and work of Rudolf Steiner.

I discussed Steiner's philosophical work, how he provided an epistemological framework for Goethe's endeavors in the natural sciences. I briefly told of Steiner's book *The Philosophy of Spiritual Activity*, in which he attempted to demonstrate, following the methods of modern science, how human beings can experience themselves as spiritual beings through the human activity of thinking.

From this foundation, a picture of humanity as an inwardly active, spiritually mobile being, Steiner built the work that led to a variety of institutions and other endeavors to create a cultural renewal.

A unification of human experience—of religion, art and science, Steiner maintained, depends on such a new, holistic picture of humanity. It is the inner, yet objective activity of thinking that can be enhanced and enlivened to enable people to develop in themselves the abilities required to deal with the forces that threaten to destroy us. I undertook to examine Steiner's original contributions to agriculture and education to determine whether they can provide insights into the problems increasingly apparent in those fields. Steiner's work provides ways of apprehending the kingdoms of nature and man's relation to them, and the means by which to educate people, to awaken the abilities of which I have spoken?

Part Two contains a sketch of agricultural history from the perspective of the evolution of human consciousness. I investigated Steiner's concepts of historical development which I then applied to agriculture. I did not intend to give a thorough treatment of the history of agriculture, but rather to examine that history as a manifestation of the development of human faculties, of the abilities by which humans have thought, felt, acted, and constructed their view of reality. This section concluded that such faculties historically were taught in the temple sanctuaries, which Steiner called Mystery centers, or Mystery schools.

Steiner held that the increasing individualization of humankind, the evolution of thinking, the receding of the old clairvoyant perceptions, and the human domination over natural forces, all stemmed from the gradually maturing human ego, the "I". Steiner considered the incarnation of Christ as the physical embodiment of the birth of the Logos in the soul, which birth Steiner said had been the focus of the old Mysteries—the penetration to direct spiritual perception of the Cosmic Word. I traced the development of the intellect to modern times with its ultimate fragmentation of society and knowledge, and the progressive descent into materialism. Part Two thus attempts to provide a cultural-historical background against which Steiner's work in education and agriculture may be seen.

In Part Three, I devoted separate chapters to biodynamic agriculture and Waldorf education, the names by which Steiner's methods and principles in these fields are known. Steiner based his work in both spheres on his image—or model—of the relation of the human being to the kingdoms of nature. I tried to show Steiner's concept that problems in education and agriculture are ultimately tied to human consciousness and its evolution. Both the farmer and teacher must cultivate new abilities within themselves if they are to work according to the principles Steiner gave. Whether one farms strictly for yield and teaches to impart a certain quantity of knowledge, or whether one increases soil life and plant and animal health and gives children living images and concepts that can grow with them—these differences are a matter of *consciousness*, of how one perceives and thinks, of one's basic attitudes toward life.

Ultimately then, teaching and farming by Steiner's methods create a new relationship between humanity and the world, the knower and the known, subject and object, teacher and child, farmer and land. The farmer and teacher must enter into an active, creative collaboration with the dynamic forces that weave through the world. The new relationship is a healing one, for to enter into such a collaboration, to *one yourself* with plants, animals, or children, heals the rift originally caused by ego consciousness. Building on the self-sustaining power of an enlivened thinking, the farmer assists the living organism of his farm to grow, and the teacher helps the child's individuality to realize its potential.

As the environment becomes more and more toxic through pollution, as the social fabric grows more threadbare, both the earth and humankind need a healing influence. The farmer strives to harmonize earthly and cosmic forces. The teacher seeks to establish a rhythm in the classroom as well. The farmer knows what principles in his own being he shares with the mineral, plant, and animal kingdoms. The teacher knows the developmental sequence of the maturity of these principles in the human being and paces her lessons accordingly. Just as the farmer works with the metamorphosis of plants and of his farm organism, so the teacher helps the child's development.

The Waldorf curriculum aids in this rhythmical approach, with subjects such as history, geography, and physics given first in pictorial form, then recapitulated in later years in a more intellectual manner to coincide with the maturing intellect. For instance, from the first grade the children play musical instruments. In the sixth grade an acoustical physics block brings the element of sound in this music-making to the child's consciousness, but in a pictorial way. In high school physics, the actual mathematical laws of acoustics are studied; this study builds on the years of music *and* the observations of the sixth grade physics block, and recapitulates them, drawing the old experiences from the subconscious, and revealing them in a new way suitable for the dawning intellect.

Agriculture increasingly is called on to heal the ravaged earth, and education more and more is needed to heal, to become "curative" in dealing with hypertension, alienation, apathy, and actual violence. It is the healing aspect of biodynamics and Waldorf education that I want to emphasize in this summary: to heal the earth, to heal children and assist their souls to become healthy and strong vehicles for their individuality; and finally, to heal the teachers and farmers who practice these methods by uniting them with their work. As the farmer Carsten Pank writes: “Any attempt to create a farm organism without the farmer being able to develop a personal relationship to his work with nature can only result in failure.”[3]

The teachers' personal involvement with the growing children allows them to choose the appropriate story to help a child through a difficult period. Another example of the healing aspect of Waldorf education is the effort made during the ninth year, through Old Testament stories and the farming and house-building blocks. These things help to overcome the dualism and alienation that often arises at this time. This effort continues during the nature studies of the next three years.

It is through an enlivened thinking, perhaps first of all through an enlivened *imagination*, that the farmer perceives the living farm organism and the teacher perceives the evolving children. Here is the opposite of reductionism, fragmentation, and purely intellectual analysis. Here is the necessity to create inwardly a whole, a living image of a living, growing being.

Since at least the time of Descartes, the "great paradigm", in Thomas Kuhn's meaning, has been that mind somehow is different than matter. Yet Steiner asserted that the very opposite is true, in the sense that it is one's own vitality, one's life forces, that allows the perception of the living organism. The imagination of the living plant or of the farm organism is of the same quality as the farmer's life-forces. For Steiner, then, to *one yourself* with an object in imagination has a literal meaning. Here is a "back to nature" movement with a new twist, an inner merging with the kingdoms of nature, yet in full waking consciousness. Humankind today needs the same sort of faculties possessed by initiates of the ancient Mysteries, but we must take hold of them in a different way—with clear consciousness, not with dreamy clairvoyance. Steiner described this step in the development of human consciousness as the time when ". . . one's process of thinking has reached such a form that it can attain to the reality of being which is in the phenomena of nature."[4]

The situation with many people today was depicted by Rudolf Steiner years ago:

> And so large numbers of people at present are faced with work that turns them back upon themselves. The environment cannot interest them, nor what they do from morning till night, unless it be presented to them so that they *can* find it interesting. What interests them first and foremost—and this is where we must begin—is what confronts a man when he is alone with himself after work and can simply concentrate his attention on his own humanity.[5] (Steiner's emphasis)

As Dr. Franz Winkler pointed out:

> Steiner's gift to the world was a moral and meditative way to objective vision, a way appropriate to the psychological and physiological constitution of Western man. If accepted in the spirit of humility, altruism and truthfulness with which it was given, it could bridge the existing cleft between a man's religious conviction and his intellect and will. It could add comprehension to our existing knowledge and thus revive the vision without which our generation will hardly find the solution to its problems.[6]

**Evaluations and Applications**

In terms of numbers, the biodynamic and Waldorf education movements seem somewhat insignificant on a global scale, even if one restricts the area to the Western world. It is when the movements are examined from the aspect of *quality* that a more telling picture arises, for people seem increasingly concerned with the quality of their lives, a quality that has suffered from the one-sided rational materialism of the past two centuries. From this perspective, the hundreds of biodynamic farms and market gardens, the many home gardens, and the over four hundred Waldorf schools appear as models for what agriculture and education can achieve. In quantifiable results, they have had success comparable to, or greater than, their conventional counterparts.[7] Yet I believe that their true worth will become steadily apparent in the future, as more and more human beings look for meaningful alternatives. It is important to note that the work done in education and particularly agriculture following Steiner's indications is still in a process of growth and development; the idea of a closed system of thought is the very opposite of what Steiner desired.[8]

For the developing nations, these methods could provide several things. The mixed farm approach could allow farmers to circumvent the cash-crop syndrome of most of these nations. Since mixed crop operations often have a more regional marketing approach, viable regional associations of consumers and growers could arise. Indeed, the growing number of "community-supported" market gardens has, in the past few years, brought many new people into a renewed relationship with agriculture. Certainly the biodynamic principles offer a desirable alternative to the high-technology methods of the "Green Revolution", which are feasible only in highly mechanized, intensively capitalized chemical farming situations. Biodynamics is applicable to the small farms and gardens that are typical of native populations, such as the Indians in Guatemala and the Kikuyu and Luo tribes in Kenya. The Rural Development program of Emerson College has established a pilot program in the Dominican Republic. The philosophy of our relation to nature inherent in the biodynamic approach is much more palatable to such people than the scientific-technological Green Revolution outlook.

Waldorf education will allow a creative approach to the sensitive problems that have arisen as western scientific culture penetrates into the more "primitive" cultures. For example, the story curriculum could include fairy tales, folk tales, and myths of the native peoples. Children could gain through pictorial forms a feeling for the universal image of the human being, and how the image of man of his own culture compares with many other cultures.

Colin Turnbull writes of the African soul torn between the old tribal consciousness and the westernized culture, and being accepted by neither one:

> In both cases there is a void in the life of the African, a spiritual emptiness, divorced as he is from each world, torn in both directions. To go forward is to abandon the past in which the roots of his being have their nourishment; to go backward is to cut himself off from the future, for there is no doubt about where the future lies.[9]

He continues: "The African has been taught to abandon his old ways, yet he is not accepted in the new world even when he has mastered its ways. There seems to be no bridge, and this is the source of his terrible loneliness."[10]

Turnbull's comments remind me of Steiner's characterization of the old pictorial clairvoyance receding into dream consciousness as the intellect began to grow. Is it not possible that the African experience (and those of many Third World areas) is a telescoped version of this process? What once took place over many centuries today takes place in a single lifetime.

Many people in the West today find it similarly hard to relate to the scientific-technological culture and rebel against it in many ways, ranging from violent acts to drug-induced oblivion. Rudolf Steiner anticipated such a situation:

> . . . the spirit however will not let itself be suppressed. Institutions which try to regulate education merely from the point of view of the economic life would be an attempt toward such a suppression. This would cause the free spirit of man to be in a constant state of revolt, welling up from its very depths. Constant disruption of the social structure would be the

unavoidable consequence of an arrangement which tried to organize the social system on the same basis as the processes of production.[11]

The spiritual emptiness of which Turnbull speaks, and the constant revolt referred to by Steiner derive from a culture which can no longer provide a nourishing world-conception, an image of humanity and its place in the world process. In this book, I have examined Rudolf Steiner's contributions to education and agriculture which he based on a world-conception that strives to show humankind the way to a new, conscious citizenship in the universe, and out of which people are working to heal the earth and to help children to grow into a healthy maturity. The biodynamic agriculture movement, the Waldorf education movement, and the adult institutions which provide educational opportunities to study anthroposophical thought—these endeavors provide meaningful alternatives for those people who search for another way to live and work.

**Some Reflections and Possible Areas for Further Research**

As I look back over this study of agriculture and education, I will attempt to state several questions which present themselves, along with some evaluations derived from my work.

*1. Do these educational and agricultural ideas have any validity in a non-anthroposophical setting, or are they sectarian in nature?*

Here we can recall E. E. Pfeiffer's statement that people who apply Steiner's agricultural ideas are led into a new attitude toward nature, a "creative collaboration" with natural forces, through the application of the biodynamic principles. Carlgren, Edmunds, and Harwood point out in their books that children are never taught anthroposophical ideas in Waldorf schools; yet they agree that teachers who work in Waldorf schools generally pursue Steiner's concepts of child development and psychology with deep interest. It seems reasonable to conclude that a certain basic attitude of mind is necessary even to look for alternatives to conventional farming or education. Once the desire for an alternative approach is present, then Steiner's methods and concepts can provide new perspectives for the teacher or the farmer. My conversations with teachers and farmers, and the literature of the Waldorf education and biodynamic agriculture movement have revealed that this dissatisfaction with conventional approaches nearly always precedes the search for new ways.

Waldorf teachers and biodynamic farmers are by no means all committed anthroposophists. There are no statistical studies to determine what percentage of either profession are members of the Anthroposophical Society, or what percentage have studied what amount of Steiner's teachings. The only figure I found was that thirteen people of the 428 practitioners of biodynamic methods who responded to a recent questionnaire have studied in one of the available biodynamic courses. Of the 428 people, 303 responded that biodynamics is "a deep spiritual commitment to the earth and man" when asked to define it.[12]

I think that Steiner's methods are meant to provide an alternative approach, one based on a holistic, spiritual understanding of humanity and nature as a living organism. As such, they appeal to those who think that the rational, materialistic scientific world-view does not provide meaningful answers to the large questions of life, nor do they sufficiently reveal of humankind's relation to the kingdoms of nature. Steiner, Pfeiffer, Koepf, Edmunds, and Harwood agree that such people yearn for a personal connection with their work and want to see a spiritual meaning in their endeavors. Steiner himself hoped that no sectarian attitude would arise in those who carried out work based on his ideas. He said that anthroposophy

> . . . seeks rather to present ideas and concepts only in order that they may become as vital within us, on the spiritual plane, as our life's blood itself, so that man's activity, not only his thinking, is stimulated. A philosophy of life in accord with spirit thus reveals itself as a social as well as a cognitive impulse.[13]

Steiner's work was offered to all who seek to work toward a renewal of society and culture. He did not regard his efforts as a finished product, but rather as a beginning. It is certainly true that many of those who have practiced biodynamics or taught in Waldorf schools have recognized the social implications of Steiner's work and have attempted to create new social forms with which to carry out more effectively the ideals of the anthroposophical approach to education or agriculture.

*2. Does the implementation of biodynamic agricultural methods imply a total lifestyle? Can they be applied by most farmers?*

The biodynamic methods imply no total lifestyle, and they can be applied by any conventional farmer who is willing to do so. It is not the actual methods, as much as the attitude that is important. What is important, say the writers whose work I consulted—Steiner, Pfeiffer, Koepf, and Meir, is that the farmer cultivates "a personal relationship to his work with nature", that he begins to enter into that *creative collaboration* with the dynamic forces that weave through the life of the world.[14] Any farmer who desires such a working relationship, and who possesses the requisite practical skills, can apply the biodynamic principles. As Pfeiffer wrote: "One must learn to understand that there is a difference between mere application of the methods and creative collaboration."[15]

*3. Do Steiner's indications for agricultural education apply only to the problems of an elite?*

Problems of soil erosion and depletion, poor nutrition, water pollution, pesticide and herbicide build-up in the biosphere, rural re-population, cultural renewal—these problems do not refer to an elite group or any other special group. A growing number of writers, to some of whose work I have referred in this study, assert that the solution of these problems is bound to the destiny of the planet. it is particularly among young people that one can see the need for a new agricultural-educational approach. Many people fled from the evils of "civilization" to seek solace in the quiet groves of "nature" to live off the land, only to find that it's not so easily done. For example, my grandfather, a lifelong rancher, often pointed out that agricultural work demands a wide range of skills yet usually pays only unskilled wages. Life in rural settings requires many skills and attitudes which schools today, and personal feelings for nature, no matter how sincerely felt, do not provide. Ironically, Rudolf Steiner addressed himself to these very problems in a lecture to young people given at Koberwitz when he gave the agricultural course. He spoke of the life of the peasantry and how different the instinctive certainty and deep insight into nature of these earthy people was from the abstractions of natural science and modern philosophy, "from Natural Science and the so-called civilization which have become so remote from all true existence."[16] Steiner described the intense yearning in the hearts of his audience for a conscious rediscovery of the knowledge of the spirit in nature.

*4. Is biodynamic agriculture an economically viable method of farming?*

According to evidence presented by Koepf, Pfeiffer, and Pank, biodynamic farming is increasingly viable economically due to the following factors: (1) The quality and quantity of animal fodder raised on the farm means little has to be brought in. (2) Few veterinary-medical bills are incurred, which is a result of the nutritive quality of farm-raised fodder and because the animals are not pushed toward over-production. (3) The farms escape the high cost of fertilizer, pesticides and herbicides. (4) Mixed-crop farming allows a variety of produce and often regional or local marketing.

*5. What does experimental research say about the scientific validity of biodynamic agriculture?*

In my chapter six, I presented a short synopsis of the research done in biodynamics. Readers are referred to the work of Koepf and Pfeiffer, and to the four thousand pages of the *Biodynamics* quarterly for the many examples of scientific studies that have been done. Of particular benefit is chapter nine: "Quality Through Growing Methods" in Koepf, Pettersson, and Schaumann's *Bio-dynamic Agriculture,* and many examples of experimental data are contained in H.H. Koepf's *The Biodynamic Farm*. Perhaps the key contributions of biodynamic research have been to find ways of quantifying qualitative factors, and in determining the complex interrelationship of factors that produce quality produce. Koepf writes that: "Work on both the thought content of the concept of quality and on methods to determine quality is unceasing."[17]

The following words of H. H. Koepf will sum up questions four and five:

> One of the recurring questions is: Quantity or quality? The biodynamic grower is interested in the first place in quality. The economic viability of his work, however, also depends on quantity. Since biodynamic work started in the twenties, the demands on field and stable in terms of yields have approximately doubled. Biodynamic producers are faced with the problem of remaining competitive despite their different attitudes and aims. Yields from biodynamic farms have risen at approximately the same rate as those on other farms. The ever recurring question is: Can we increase the yields still further and yet combine this with the quality

achieved hitherto? In essentials, research over the past fifteen years and also on-going analysis of products in practice have found that this question can be answered in the affirmative. The measures that work positively toward quality are better known today than before, particularly in the degree toward which they work. To apply them at the right time in the right way in different places is one of the most important tasks of farm management.[20]

*6. What are the implications of teaching and employing these methods for under-developed nations? What are the social, economic and psychological ramifications?*

I attempted to outline the social and psychological ramifications for teaching and employing biodynamic methods in this chapter.

My hypothesis that the changes of consciousness described by Steiner take place in peoples relatively recently exposed to western civilization needs to be researched further. There are several projects and programs that have worked well in developing countries. Interested readers can find many articles in *Biodynamics* that describe some of these efforts.[19]

Further research is needed in several other areas. for example, there have been no studies to show to what extent Waldorf-trained people have a greater appreciation for biologically safe agriculture or a holistic approach to nature than other people. No research has undertaken to discover and compare the degree of dedication to humanity and to the earth of graduates of conventional agricultural schools and biodynamic agricultural programs.

In conclusion, I have come to see that agricultural education and practical training should, wherever possible, not stand alone as a scientific, academic speciality; rather, they should be integrated with a holistic approach to culture within agricultural schools. This approach should include a rigorous critique of the modern scientific world-view—its strengths and its limitations.

The work of Rudolf Steiner in education and agriculture, and the movement for cultural renewal that has arisen from Steiner's work, presents a holistic alternative to contemporary practices. Other such alternatives, based on spiritual insight and a philosophy of humanity and its place in nature, and that come to terms with modern science, may exist. I have not found any that take the comprehensive view of Steiner and those who have followed him. It would be well to add once again, and as a final thought, that while Steiner's conclusions and indications have had a seedlike power in medicine, philosophy, education, architecture, art and agriculture, the work now being done is by no means a finished product. Rather, every effort is made to build on Steiner's contributions to enliven these areas of our culture.

**Footnotes for Chapter Nine**

I. Rudolf Steiner, *Human Values in Education* (London:1971) 144-145.
2. von Helmholtz quoted in Henry Williams, "The Doctor and the Farmer," *Biodynamics*,

113(Winter 1975) 7.
3. Carsten Pank, "Dirt Farmer's Dialogue," *Biodynamics*, 105 (Winter 1973) 31.
4. Rudolf Steiner, *The Course of My Life* (New York:1951) 24.
5. Rudolf Steiner, *The Tension Between East and West* (London: Stodder and Houghton, 1963) 152.
6. Franz Winkler, *Man: The Bridge Between Two Worlds* (Blauvelt, New York: Rudolf Steiner Publications, 1954) 25.
7. See Koepf, Pettersson, and Schaumann, *Biodynamic Agriculture* (Mew York: 1977) and Frans Carlgren, *Education Towards Freedom* (East Grinstead: Lanthorn Press, 1973).
8. Koepf, Petersson, and Schaumann, lbid., 400, 201.
9. Colin Turnbull, *The Lonely African* (New York: Simon and Schuster, 1962) 15.
10. Ibid., 16.
11. Rudolf Steiner quoted in Carlgren, *Education Towards Freedom*, 109.
12. Maria C. Linder, "Report on the Questionnaire," *Biodynamics*, 120(Fall 1976) 40.
13. Steiner, *The Tension Between East and West*, 114.
14. Ibid., 110; E.E. Pfeiffer, *Biodynamics* (Springfield, Ill.: Biodynamic Farming and Gardening Assn., 1948, 1956) 30.
15. Pfeiffer, Ibid.
16 Steiner, *Youth's Search in Nature* (Spring Valley, N.Y.: Mercury Press, 1975) 4.
17. Koepf, Pettersson, and Schaumann, *Biodynamic Agriculture* (Spring Valley, New York: 1979) 386.
18. Ibid., 386-387.
19. For example: Barbara Booth, "Kimberton Hills—An International Perspective," *Biodynamics*, 160 (Fall 1986) 6; Gregory Booth, "Teaching Alternative Agriculture," Ibid., 7-10; Barbara Booth, "Tamil Nadu Theological Seminary and the Rural Theological Institute, *Biodynamics*, 159(Summer 1986) 13-18; Chris Stearn and Chen Li Ju, "Permanent Composting Beds for Intensive Vegetable Production in Taiwan," *Biodynamics*, 159(Summer 1986) 53-61; and particularly "Biodynamic Training in the Dominican Republic," *Biodynamics*, 162( Spring 1987) 31-41.

## Appendix A: Rudolf Steiner's Basic Works on Christianity

This appendix owes much to the bibliographic essay in Stewart C. Easton's *Man and World in the Light of Anthroposophy* (New York: 1975) 214-216.

1. Introduction and background:

Rudolf Steiner, *Christianity as Mystical Fact*, 2nd. ed. (New York: 1972).

*The Spiritual Guidance of Man* (New York: 1970).

*Building Stones for an Understanding of the Mystery of Golgotha* (London: 1972).

2. Of central importance, and given here in the chronological order in which they were delivered, since Steiner built on the information he had imparted before:

Rudolf Steiner, *The Gospel of St. John*, 12 lectures given in Hamburg, May 1908 (New York: 1973).

*The Gospel of St. John in its Relation to the Other Three Gospels*, 14 lectures given in Cassel, June-July 1909 (New York: 1948).

*The Gospel of St. Luke*, ten lectures given in Basel, September 1909, 3rd ed. (New York: 1975).

*Deeper Secrets of Human History in the Light of The Gospel of St. Matthew*, twelve lectures given in Berlin, November 1909 (London: 1957).

*The Gospel of St. Matthew*, twelve lectures given in Bern, September 1910 (London: 1946).

*Background to the Gospel of St. Mark*, ten lectures given at intervals in Berlin, October 1910-June 1911 (London: 1970).

*The Gospel of St. Mark*, ten lectures given in Basel, September 1912 (New York: 1950).

3. Important as a supplement to the above are:

Rudolf Steiner, *The Christ Impulse and the Development of the Ego Consciousness*, seven lectures given in Berlin, October 1909-May 1910 (New York: 1975).

*Esoteric Christianity and the Mission of Christian Rosenkreutz*, lectures given in various cities, 1911-1912, 2nd ed. rev. (London: 1984).

*The Fifth Gospel*, seven lectures given in Oslo and Cologne, October-December 1913, 2nd ed. rev. (London: 1968).

*From Jesus to Christ,* ten lectures given in Karlsruhe, October 1911 (London: 1973).

*The New Spirituality and the Christ Experience of the 20th Century,* seven lectures given in Dornach, October 1920 (London: 1988)

*The Universal Human,* four lectures given in Munich and Bern, 1909-1916 (New York: 1990).

Interested readers can find many books by other anthroposophical writers on esoteric Christianity by consulting the catalog of the Anthroposophic Press, RR 4, Box 94AI, Hudson, NY 12534, (518) 851-2054.

## Appendix B: Addresses of Places to Study Biodynamic Agriculture and Waldorf Education

I. *Foundation year courses in Steiner's work.*

1. Emerson College, Forest Row, Sussex RHI8 5JU, England.

2. Waldorf Institute, 260 Hungry Hollow Rd., Spring Valley, NY 10977.

3. Rudolf Steiner College, 9200 Fair Oaks Blvd., Fair Oaks, CA 95628.

4. Waldorf Institute of Southern California 17100 Superior Street, Northridge, California 91324.

II. *Training Courses in Biodynamics.*

1. Emerson College, England. For the Biodynamic Farming Course and the Center for Rural Development.

2. Training Course, Kimberton Hills Camphill Village, P.O. Box 155, Kimberton, PA 19442.

3. Christoph Meier, Hawthorne Valley Farm, RD 2 Harlemville, Ghent, Ny 12075.

4. Warmonderhof School for Biodynamic Agriculture, Thedingsweert 3, NL 2720 Kerk-Avezaath, Holland.

5. Biodynamic Farming and Gardening Association, P.O. Box 306, Napier, New Zealand. One year course registered with the government.

6. Instituto Biodinámico, Rua Amando de Barroa 1831, 18610 Botucatu SI, Brazil.

7. Biodynamic Farming and Gardening, Skilleholm, 15300 Järna, Sweden. Two year farming course.

8. Landbauschule Dottenfelderhof, D6368 Bad Vilbel, Germany. One year course in biodynamic agriculture, with admission following four years of work in horticulture or agriculture.

9. Ecole d'Agriculture Biodynamique, L'Ormoy, Saint Laurent, F18330 Neuvy sur Barangeon, France. Two year course.

10. Rural Development Service Group, 923 Tilden St., Las Vegas NM 87701.

III. *Waldorf Education teacher-training courses.*

1. Waldorf Institute, 260 Hungry Hollow Road, Spring Valley, NY 10977 (914-425-0055). Has an early childhood program, class teacher program, program for

school administration, and a curative education program.

2. Emerson College: a teacher training program.

3. Rudolf Steiner College, 9200 Fair Oaks Blvd., Fair Oaks, CA 95628 (916-961-8727). Has an early childhood program, teacher training program, and an arts program.

5. Waldorf Institute of Southern California, 17100 Superior Street, Northridge, California 91324.

IV. *Biodynamic Conferences and Lectures*

1. Biodynamic Farming and Gardening Association, P.O. Box 550, Kimberton, PA 19442 (215-935-7797).

2. Regional and national conferences are announced in the *Biodynamics* quarterly, available from the Biodynamic Association, and in the Association's newsletter.

V. *Apprenticeships.*

Contact the Biodynamic Farming and Gardening Association, P.O. Box 550, Kimberton, PA 19442 (215-935-7797).

## SELECTED BIBLIOGRAPHY

Bibliographical Note

For Rudolf Steiner's books on Christianity, see Appendix A.

In the bibliography, as in the footnotes as the end of each chapter, I have cited the books published by the two major firms that publish anthroposophical authors in the following manner:

1. Books published by the Anthroposophic Press, Spring Valley and Hudson, New York, are cited as (New York: date).
2. Books published by the Rudolf Steiner Press, London, are cited as (London: date).

### I. Articles


Adams, George. "Space and Counter-space." in Harwood, ed. *The Faithful Thinker* (London: 1961) 123-140.

*Biodynamics,* 160(Fall 1986). The entire issue is devoted to training for work in biodynamic agriculture.

Booth, Barbara. "Germ Plasm: Slave Trade of the 20th Century," *Biodynamics,* 181(Winter 1991-92) 2-14.

Brandenburg, David. "Commentary on Eighteenth Century British Agriculture." *Agricultural History,* 42(1974) 19-24.

Clary, Mike. "The Dilemma of Michigan's Family Farms." *Parade* (26 November 1976) 26-36.

Davy, John. "Rudolf Steiner—Initiate of the Will." in Davy, ed. *Work Arising from the Life of Rudolf Steiner* (London: 1975) 1-23.

Edwards, I. E. S. "The Early Dynastic Period in Egypt." *Cambridge Ancient History.* I(London:Cambridge University Press, 1969) 45-68.

Gadd, C. J. "The Cities of Babylon." *Cambridge Ancient History* 1,2(London: Cambridge University Press, 1969) 121-137.

Gregg, Elizabeth Speiden. "The Early Days of Biodynamics in America." *Biodynamics* 119(Summer 1976) 25-39; 120(Fall 1976) 7-21; and 121(Spring 1977) 16-23.

Grotzke, Heinz. "Rudolf Steiner's Impulse to Herbology." *Biodynamics.* 90(Spring 1969) 26-33.

Hahn, Herbert. "The Birth of the Waldorf School from the Threefold Social Movement." *Golden Blade* (1958) 50-70.

Harwood, A. C. "Threefold Man." in Davy, ed. *Work Arising from the Life of Rudolf Steiner* (London: 1975) 27-39.

Hauk, Günther. "Gardening as a Subject in the Waldorf School," *Biodynamics*, 169 (Winter 1988-1989) 21-28.

Hutchins, Eileen. "The Normal Child." in Davy, ed. *Work Arising from the Life of Rudolf Steiner* (London: 1975) 77-92.

Isaac, Erich. "On the Domestication of Cattle." in Shepard and McKinley, eds. *The Subversive Science: Toward an Ecology of Man* (Boston: Houghton Miflin, (1969) 191-207.

Jeetze, Hartmut von. "Agriculture and the Camphill Movement," *Biodynamics* 114(Spring 1975) 1-6.

"Biodynamic Relations Between the Farmer and the Land." *Biodynamics* 115(Summer 1975) 10-15.

Koepf, H. H. "Three Lectures on Biodynamics." *Biodynamics.* 88(Fall 1968) 1-50.

Kroobner, Richard. "The Settlement and Colonization of Europe." *Cambridge Economic History of Europe* (Tendon: Cambridge University Press, 1966) 72-112.

Lindenberg, Christof-Andreas. "Thoughts on Curative Education and Social Therapy," *Journal for Anthroposophy*, 44(Winter 1986) 37-44.

Linder, Maria C. "A Review of the Evidence of Food Quality." *Biodynamics.* 107(Summer 1973) 1-12.

Linkhachev, B. "A National Seed Bank," *Biodynamics.* 121(Winter 1977) 31-33.

Maguire, Dan. "A Critical View of Techno-industrial Agriculture." *Biodynamics.* 107(Summer 1973) 13-36.

McGavin, Janet. "Beginning of the Camphill Work in America." *Journal for Anthroposophy*, 44(Winter 1986) 34-36.

Meir, C. A. "What Is a Farm?" in Harwood, ed. *The Faithful Thinker* (London: 1961) 221-232.

Mingay, G. E. "The Agricultural Revolution in English History: A Reappraisal." *Agricultural History.* 26(1963) 123-133.

Moore, Hilmar. "The Living Earth," *Biodynamics*, 170(Spring,1989)20-29.

Morrow, Joel. "The Covelo Farm School," *Biodynamics*, 169(Winter 1988-89) 4-5.

"A Thread from the Tapestry Alanus Wove: Nature and Inner Development in Alan of Lille and Bernardus Silvestris," *Journal for Anthroposophy*, 51(Fall 1990) 5-24.

Pank, Carsten J. "A Letter to Those Who Look for Employment on a Biodynamic Farm." *Biodynamics*, 118(Spring 1976) 35-36.

Parain, Charles. "The Evolution of Agricultural Techniques in the Middle Ages." *Cambridge Economic History of Europe*. I(London: Cambridge University Press, 1966) 130-164.

Pietzner, Cornelius. "Camphill: Celebrating 25 Years in America." *Journal for Anthroposophy*, 44(Winter 1986) 29-31.

Pfeiffer, E. E. "New Directions in Agriculture."*Golden Blade* (1958) 105-124.

Philbrick, Helen. "Biodynamics in Recent Years." *Biodynamics* 122(Summer 1977) 20-23.

Querido, René. "The Cathedral and the Great Masters of Chartres," *Journal for Anthroposophy*, 45(1987) 47-65 and 46(1987) 5-20.

Rotheraine, L.A. "The Hibernian Mysteries as a Foundation of Biodynamics," *Biodynamics,* 181(Winter, 1991-92) 35-38.

Schmidt, Gerhardt. "Aspects of Protein Nutrition." *Biodynamics,* 120(Fall 1976) 1-6.

Smith, Sidney. "Senacherib and Escarhaddon." *Cambridge Ancient History*. II(London: Cambridge University Press, 1970)

Steiner, Rudolf. "A Chapter in Occult History." *Anthroposophical Quarterly* (Spring 1968) 1-15.

"European Mysteries and their Initiates." *Anthroposophical Quarterly* (Spring 1974) 165-173,

"Natural Science and its Boundaries." *Golden Blade* (1962)) 1-13.

Thompson, R. Campbell. "The Influence of Babylon." *Cambridge Ancient History* III(London: Cambridge University Press, 1971) 235-252.

White, Lynn Jr. "The Roots of Our Ecological Crisis." *Science*. 155(July 1963) 1207-1238.

Williams, Henry. "The Doctor and the Farmer, *Biodynamics*. 113(Winter 1975) 4-11.

Wolf, Bernard. "Agricultural Work at Camphill Special Schools," *Biodynamics,* 108(Fall, 1973).

## II. Books

Allen, Paul M. ed. *A Christian Rosenkreutz Anthology* (Blauvelt, New York: Rudolf Steiner Publications, 1968).

Bailey, Liberty Hyde. *The Holy Earth* (Ithaca: New York State College of Agriculture, 1980).

Begg, Ean. *The Cult of the Black VIrgin* (London: Arkana, 1986).

Bemmelen, D. J. van. *Zarathustra.* 2 vols.(Zeist, Netherlands: Uitgeverij Vrij Geesteslebens, 1968) reprinted by Rudolf Steiner College Publications, 1988.

Berry, Wendell. *The Unsettling of America: Culture and Agriculture* (New York: Avon Books, 1977).

Bock, Emil. *The Three Years: The Life of Christ Between Baptism and Ascension* (London: Christian Community Press, 1955).

Braudel, Fernand. *Capitalism and Material Life, 1400-1800* (New York: Harper and Row, 1973).

Bronowski, Jacob. *The Ascent of Man* (Boston: Little, Brown, 1973).

Carlgren, Frans. *Education Towards Freedom: Rudolf Steiner Education—a Survey of Waldorf Schools Throughout the World* (East Grinstead; Sussex: Lanthorn Press, 1973).

Carter, Vernon Gill and Dale, Tom. *Topsoil and Civilization.* Rev. ed.(Norman: University of Oklahoma Press, 1974).

Commoner, Barry. *The Closing Circle: Nature, Man, and Technology* (New York: Alfred A. Knopf, 1972).

Davy, John, ed. *Work Arising from the Life of Rudolf Steiner* (London: 1975).

Dubos, René. *The Dreams of Reason: Science and Utopias* (New York: Columbia University Press, 1961).

Easton, Stewart C. *Man and World in the Light of Anthroposophy* (New York: 1975).

*Rudolf Steiner: Herald of a New Epoch* (New York: 1982).

Edmunds, Francis. *Rudolf Steiner's Gift to Education: The Waldorf Schools.* 3rd ed. rev.(London: 1975).

Frankfort, Henri and others. *Before Philosophy: The Intellectual Adventure of Ancient Man* (New York: Penguin Books, 1949).

Fussell, G. E. *Farming Techniques from Prehistoric to Modern Times* (Oxford: Pergamon Press, 1965).

*The Classical Tradition in Western European Farming* (Rutherford, New Jersey: Fairleigh Dickinson Press, 1972).

Gardner, John. *The Waldorf Approach to Education: Questions and Answers* (Sacramento: Sacramento Waldorf School, n.d.).

Glas, Werner. *Speech Education in the Primary Grades of Waldorf Schools* (Wilmington, Del.: Sunbridge College Press, T974).

*The Waldorf Approach to History.* 2nd ed. rev.(New York: 1985)

Goethe, Johann Wolfgang von. *The Metamorphosis of Plants* (Springfield, Ill.: Biodynamic Farming and Gardening Assn., 1974).

Gottschalk, Louis, Mckinney, Loren C. and Pritchard, Earl C. *The Foundations of the Modern World.* 2 vols.(London: George Allen & Unwin, 1969).

Grigg, D. B. *The Agricultural Systems of the World: An Evolutionary Approach* (London: Cambridge University Press, 1974).

Hall, Ross Hume. *Food for Nought: The Decline in Nutrition* (New York: Vantage Books, 1976).

Harwood, A. C., ed. *The Faithful Thinker: Centenary Essays on the Work and Thought of Rudolf Steiner* (London: Hodder and Stoughton, 1961).

*The Way of A Child.* 4th ed. rev.(London: 1967).

Hawkes, Jacquetta and Wooley, Sir Leonard. *Prehistory and the Beginnings of Civilization* (London: George Allen & Unwin, 1967).

*The First Great Civilizations* (New York: Alfred A. Knopf, 1973).

Hemleben, Johannes. *Rudolf Steiner: A Documentary Biography* (East Grinstead, Sussex: Henry Goulden Ltd., 1975).

Hyams, Edward. *Soil and Civilization* (New York: Harper Colophon Books, 1973).

Jones, E. L. *Agriculture and the Industrial Revolution* (New York: Halstead Press, 1974).

Kerridge, Eric. *The Agricultural Revolution* (London: George Allen & Unwin, 1967).

*The Farmers of Old England* (Tobowa, New Jersey: Rowman and Littlefield, 1973).

Koepf, H. H., Pettersson, Bo, and Schaumann, Wolfgang. *Biodynamic Agriculture: An Introduction* (New York: 1976).

*Ehrenfried Pfeiffer: Pioneer in Agriculture and Natural Sciences* (Kimberton: Biodynamic Farming and Gardening Assn., 1991).

*The Biodynamic Farm* (New York: 1989)

*What Is Biodynamic Agriculture?* (Springfield, Ill: Biodynamic Farming and Gardening Assn., 1976).

Lehrs, Ernst. *Man or Matter: An Introduction to a Spiritual Understanding of Nature on the Basis of Goethe's Training of Observation and Thought.* 2nd ed. rev.(London: Faber and Faber, 1958).

Merry, Eleanor C. *The Flaming Door: A Preliminary Study of the Celtic Folk-soul by Means of Legends and Myths,* rev. ed.(East Grinstead: New Knowledge Books, 1962).

Murphy, Christine, ed. *Emil Molt and the Beginnings of the Waldorf School Movement* (Edinburgh: Floris Books, 1991).

Pank, C. J. *Dirt-Farmer's Dialogue* (Sprakers, New York: B-D Press, 1976).

Pareti, Luigi, Brezzi, Paolo, and Petech, Luciano. *The Ancient World.* 2 vols. (London: George Allen & Unwin, 1965).

Pfeiffer, E. E. *Biodynamics: Three Introductory Articles* (Springfield, Ill.: Biodynamic Farming and Gardening Assn., 1948, 1956).

*Soil Fertility. Renewal, and Preservation: Biodynamic Farming and Gardening* (East Grinstead: Lanthorn Press, 1983).

Pfeiffer, Martin. *The Agricultural Individuality: A Picture of the Human Being* (Kimberton: Biodynamic Farming and Gardening Assn., 1990).

Philbrick, Helen and Gregg, Richard. *Companion Plants and How to Use Them* (Old Greenwich, Conn.: Devin-Adair, 1966).

Pirenne, Henri. *The Economic and Social History of Medieval Europe* (New York: Harvest Books, 1937).

Poppelbaum, Hermann. *Man and Animal: Their Essential Difference.* 2nd ed.(London: 1960).

*A New Zoology* (Dornach, Switzerland: Philosophic-Anthroposophic Press, 1961).

Querido, René. *The Golden Age of Chartres: The Teachings of the Mystery School and the Eternal Feminine* (New York: 1989).

*The Mystery of the Holy Grail: A Modern Path of Initiation* (Fair Oaks: Rudolf Steiner College Publications, 1992).

Querido, René and Moore, Hilmar. *Behold, I Make All Things New: Toward a World Pentecost* (Fair Oaks: Rudolf Steiner College Publications, 1990).

Roszak, Theodore. *The Unfinished Animal* (New York: Harper and Row, 1975).

Shepard, A.P. *A Scientist of the Invisible* (London: Hodder and Stoughton, 1954).

Shepard, Paul. *Man in the Landscape: An Historic View of the Aesthetics of Nature* (New York: Alfred A. Knopf, 1967).

Steiner, Rudolf. *Agriculture.* 3rd ed.(London: Biodynamic Agricultural Assn., 1974).

*A Modern Art of Education.* 3rd ed. rev.(London: 1972).

*An Outline of Occult Science* (New York: 1972).

*Background to the Gospel of St. Mark* (London: 1970).

*Human Values in Education* (London: 1971).

*Knowledge of the Higher Worlds and its Attainment.* 3rd ed. (New York: 1947)

*Man: Hieroglyph of the Universe* (London: 1972).

*Mystery Knowledge and Mystery Centres.* 2nd ed.(London:1973).

*Mysteries of Antiquity and Christianity as Mystical Fact* (Blauvelt, New York: Rudolf Steiner Publications, 1963).

*Mysteries of the East and of Christianity.* 2nd ed. (London: 1972).

*Occult History: Historical Personalities and Events in the Light of Spiritual Science* (London: Anthroposophical Publishing Co., 1957).

*Rosicrucianism and Modern Initiation.* 2nd ed. rev. (London: 1965).

*The Course of My Life* (New York: 1951).

*The Education of the Child in the Light of Anthroposophy.* 2nd ed.(London: 1975).

*The Kingdom of Childhood* (London: 1964).

*Theosophy: An Introduction to Supersensible Knowledge of the World* (New York: 1971).

*The Philosophy of Freedom* (London: 1964).

*The Roots of Education* (London: 1968).

*The Study of Man* (London: 1966).

*The Tension Between East and West* (London: Hodder and Stoughton, 1963).

*Turning Points in Spiritual History* (London: Anthroposophical Publishing Co., 1934).

*Von Seelenratseln.* trans. and ed. by Barfield, Owen as*The Case for Anthroposophy* (London: 1970).

*World History in the Light of Anthroposophy* (London: 1930).

*Youth's Search in Nature* (Spring Valley, New York: Mercury Press, 1975).

Stockmeyer, E. A. Karl. *Rudolf Steiner's Curriculum for Waldorf Schools* (Stuttgart: Verlag Freies Geistesleben, 1973).

Streit, Jakob. *Sun and Cross: The development from megalithic culture to early Christianity in Ireland* (Edinburgh: Floris Books, 1984).

Tannahill, Reay. *Food in History* (New York: Stein and Day, 1973).

Toynbee, Arnold. *Mankind and Mother Earth* (New York: Oxford University Press, 1976).

Wachsmuth, Guenther. *The Evolution of Mankind* (Dornach, Switzerland: Philosophic-Anthroposophic Press, 1961).

*The Life and Work of Rudolf Steiner* (New York: Whittier Books, 1955).

White, K. D. *Roman Agriculture* (Ithaca: Cornell University Press, 1970).

White, Lynn, Jr. *Medieval Technology and Social Change* (London: Oxford University Press, 1962).

Yates, Frances A. *Giordano Bruno and the Hermetic Tradition* (New York: Vintage Books, 1964).

*The Occult Philosophy in the Elizabethan Age* (London: Routledge and Kegan Paul, 1979).

*The Rosicrucian Enlightenment* (London: Routledge & Kegan Paul, 1972).

Zeylmans van Emmichoven, F. W. *The Reality in Which We Live* (East Grinstead, Sussex: New Knowledge Books, 1967).